The Culture of American College Radio

The Culture of American College Radio

Samuel J. Sauls

Iowa State University Press

Ames

SAMUEL J. SAULS, Ph.D., is an assistant professor in the Department of Radio, Television, and Film at the University of North Texas, Denton. Dr. Sauls has a combined total of 15 years' experience in commercial and noncommercial radio, including 10 years as the station manager for a major market university radio station. He worked for four years in Saudi Arabia as the Audio Production/Language Laboratory manager at the National Center for Financial and Economic Information in Riyadh, and was director of field operations for City Colleges of Chicago in Europe while living in Germany for two years. He serves as president elect and membership committee chair of the Texas Association of Broadcast Educators and is currently the editor of the Broadcast Education Association Membership Directory. His research interests include noncommercial broadcasting, survey research, and media pedagogy.

Iowa State University Press
2121 South State Avenue, Ames, Iowa 50014

Orders: 1-800-862-6657
Office: 1-515-292-0140
Fax: 1-515-292-3348
Web site: www.isupress.edu

♾ Printed on acid-free paper in the United States of America

First edition, 2000

Library of Congress Cataloging-in-Publication Data

Sauls, Samuel J.
   The culture of American college radio / Samuel J. Sauls—1st ed.
      p. cm.
   Includes bibliographical references and index.
   ISBN 0-8138-2068-5
   1. College radio stations—United States. I. Title.

   PN 1991.67.C64 S38 2000
   384.54'53—dc21   00-021565

The last digit is the print number: 9 8 7 6 5 4 3 2 1

*To Aunt Connie, because she always
wanted a book dedicated to her.*

*To my junior college teacher and friend,
Chuck Wright, who gave me the
motivation to believe in myself.*

*To my parents, who always supported me
in every way.*

*And to my wife, Donna, whose
encouragement, nurturing, and love
continues to inspire me.*

Contents

7   *Who's Paying for College Radio?*   *121*

8   *The College Radio Station and the Community*   *145*

9   *The Future of College Radio*   *155*

Acknowledgments

I am grateful to Craig Gill, now at the University Press of Mississippi, whose interest in my research on college radio spawned the original manuscript for this book.

The support given by my department at the University of North Texas was invaluable. Particularly, I thank my former chair John B. Kuiper, and current chair Steve Craig. Administrative assistant, Vicki Kirkley, was supportive, both in clerical assistance and helping to maintain my coherence during the publishing process.

In the initial research on the book, former graduate students Gilbert Castillo and Katherine McCullough, and undergraduate Suzy Wager, served as research assistants. Kelli McKinney, who served as my final graduate research assistant on the manuscript, deserves a great deal of praise.

I would like to recognize Frank Chorba, editor of the *Journal of Radio Studies*, and author Michael Keith, who both encouraged me to pursue this manuscript to publication.

The following helped to make contributions found throughout the manuscript. To these individuals, I am most appreciative:

Jane Williams, Rules Service Company
Fritz Kass, Intercollegiate Broadcasting System, Inc.
Russ Campbell, KNTU-FM, University of North Texas
Maryjo Adams Cochran, Sam Houston State University
Linda Dominic, National Public Radio
Robert Hall, ABC Radio Networks
Matthew L. Leibowitz, Leibowitz & Associates, P.A.
Sharon Rae, Imas Publishing Group
Rob McKenzie and Jason Fiore, WESS-FM, East Stroudsburg University
Barbara L. Miller, Strategic Media Research
Patricia Lyons and Ken A. Stein, KERA/KDTN, Dallas
Mike Elliott, Elliott Broadcast Services
Lee Martin, Radio Netherlands
John Devecka, LPB, Inc.

Ron Spielberger, College Media Advisers
Nick DeNinno, Burly Bear Network
Phil Burger, KNPR-FM, Nevada Public Radio Corporation
Robert Haber, College Media Inc.
George Gimarc, Reel George Productions, Inc.
Paula A. Jameson, Arter & Hadden LLP
Steve Klinenberg, Rare Medium, Inc.
Cary S. Tepper, Booth, Freret, Imlay & Tepper, P.C.
Craig A. Stark, KSHU-FM, Sam Houston State University
Michael Black, WEOS-FM, Hobart & William Smith Colleges
F. Leslie Smith, University of Florida
Dick Kunkel, KPBX/KSFC - Spokane Public Radio
John Pearson and John McEvoy, Minnesota Public Radio
Randall Davidson, Wisconsin Public Radio
    and to the countless individuals whose electronic messages appear
    throughout this book, but remain anonymous.

Additionally, Holly Kruse and Chuck Bailey provided literary input to the manuscript.

Finally, I would be negligent if I did not thank Judi Brown, my acquisitions editor at Iowa State University Press. Her belief in me and my work provides this book.

And, of course, I am grateful to my editor, Anne Bolen.

The Culture of American College Radio

1

Introduction: College Radio and the College Culture

I think [college radio] is a vital and necessary part of radio today! The college scene makes us all examine ourselves as to where we are. ... What defines the norm if you don't define the edges?

—George Gimarc, Author of *Punk Diary: 1970–1979* and *Post Punk Diary: 1980–1982*

Ever venture down to the far left of the FM radio dial? While you might be accustomed to listening to your favorite NPR station in the 88FM-91FM range, unique to the United States, you'll also find the majority of college radio stations located in this part of the broadcast band. This book is about those radio stations licensed to institutions of higher education and the unique culture they provide their listeners.

As a culture, college radio reflects the current climate on the campus. While an outlet for the student population, the college radio station also provides a rich perspective on the campus itself. The college radio station offers a true alternative to programming not commercially available or viable.

The best indicator of this trend is the programming of alternative music that reflects the diverse lifestyles of a "college culture." Additionally, the "open format" of the majority of college stations also distinguishes them from their commercial counterparts.

As you read this book, you will be introduced to what college radio stations are, how and why they are programmed differently from commercial radio stations, and who staffs and pays for their operation. As you might expect, a great deal of attention is devoted to the exploration of the noncommercial/alternative programming at college radio stations, particularly alternative music. In addition, this book addresses the relationship college stations have with their campus and outside communities, and the administration and management of such stations.

What you won't find in this book are the specifications to equip a college radio station. Technology, particularly broadcast and recording equipment, is evolving daily. The introduction of the digital medium has changed, and continues to change, radio broadcasting and media production at an alarming pace. And as with computers, what appears current today was probably outdated by a new development yesterday. Then, too, there is a great deal of discussion in broadcasting, particularly in college broadcasting, about the extent to which we should be moving into the digital world. Besides the cost of such a move, which system do you choose? Those of us in education have a duty to prepare broadcasters for the real world, which is currently transitioning from the traditional analog domain to digital. So if you are looking at equipping a college radio station, my recommendation is to contact a broadcast engineer, or an equipment representative or dealer, to determine your needs. Be aware that any technical information provided within this book can be easily outdated within five years.

WHY A BOOK ON COLLEGE RADIO?

The following is an e-mail message from a listserv on radio:

> *I'm just venting frustration because the local administration here at my college still isn't ready to start our station broadcasting. Instead, they want a survey done of lots of college radio stations around the nation so they can waste more time. So this is an urgent plea to any college radio station broadcasters: PLEASE help us by answering some questions for us. Maybe if we get this completed fast enough, the administration won't have any choice but to let us on the air. (Airwaves Radio Journal ListServ, Issue #2957, May 15, 1997)*

For someone in college radio, this really hit home! This is not a fluke! Questions such as these arise all the time. On a daily basis I talk to people already in college radio looking for help and just trying to figure out this thing called "college radio." But anyone can ascertain that most think it's just a matter of getting the equipment and putting the station on the air, and they don't realize the philosophies of programming, staffing, managing, etc. (until they're actually on the air and realize what they're up against). Even more important, they have no concept of the idea of how the station can and will impact the media scene in their community.

QUESTIONS TO BE ANSWERED

The original intent of this book was not a "how to" for college radio managers and programmers. Naturally, the prescriptive aspect (emphasizing the way that a college radio station should be organized, managed, and programmed) is present in the book, mainly within the practical applications. But this book is more an explanation of "why things are done that way" in college radio.

My intent in writing this book was also to give a more cultural view, to develop a broader intellectual base, thus placing college radio within the context of the larger population encompassing the mass cultural landscape. Fostering this idea is the book's emphasis on the institutional and economic structures behind college radio. Of course, the historical perspective of college radio enhances such a discussion.

The goal, then, has been to envision college radio in a wider social, cultural, and institutional context. In general, this objective provides a more significant contribution to the discussion of communication studies.

> *Of course, there's a need for material on running student radio stations!!!!! I don't know how many are out there, but how many would like to be out there? How many faculty members and students THINK about starting stations but don't know where to turn? I would think every radio faculty in the country would want such a book for their libraries at a minimum, and every student group or professor who has toyed with the idea would purchase a copy. (NACB ListServ, February 17, 1997)*

Although this book was proposed and written with a "general and accessible" approach, the emphasis of the manuscript eventually became threefold: First and foremost, it provides a "general overview" of college radio. This, for the basic reader, answers the question "What is college radio?" Then, after discussions with my colleagues in the academic world, it became apparent that a text was needed for use in classes involving college broadcasting, particularly radio. This book, then, can provide the needed classroom application, answering the question "What is college radio?" Finally, the "how to do it" approach was also needed. This approach fulfills the need for a source that college radio station operators, managers, and staff (including student volunteers) can refer to for general ideas and "answers" in the operation and development of their stations. Again, this objective will also help answer the question of "why we do things that way" in college radio, and will aid station operators in possibly creating that "real world" experience at their station (comparable to commercial stations).

PRACTICAL APPLICATIONS PER CHAPTER

Subject: Budgets ...
I am preparing my budget for the 1996-97 fiscal year. I would be interested in
how other managers prepare budgets, i.e., fund allocation for day-to-day oper-
ating expenses, travel, coverage of athletics, etc.
Without being too direct, what are the various operating budgets out there?
(NACB Discussion List ListServ, February 27, 1996)

While the e-mails and listserv discussions included throughout this text are
somewhat anecdotal, they do help the reader get a feel for the "practical"
dilemmas and questions faced daily by those in college radio. They are provided to help introduce and legitimately support the material addressed
throughout the book. This, then, brings to focus the "how to" and "classroom"
application of the book itself. (References to schools, stations, and individuals
are removed from the e-mail messages to maintain anonymity in personal communications. References to listservs are included as primary sources of information.)

How To

To provide the needed "how to do it" approach, at the end of each chapter you
will find a section entitled "Practical Applications." This section is set apart
within the chapters because, again, the original intent of the book was not a
"how to" but a "what is." Also, this format provides for the easy location of
practical information within a given area. From FCC regulations applicable to
college radio stations, to budgets and staffing, the practical applications are
included either to answer a question or, at least, to direct the reader to sources
of further knowledge. Then, to give the reader or user a concluding perspective
on managing the college radio station, a final "Practical Applications" section
is provided in Chapter 10.

For those who are already familiar with material on college radio, this book
will serve as an update of Brant's 1981 work entitled *The College Radio
Handbook.* Even more current works detailing the establishment and actual
operation of student radio are McCluskey's *Starting a Student/Noncommercial
Radio Station* (1998) and *Advising, Managing and Operating a Successful
Student/Noncommercial Radio Station* (1998). I also highly recommend that
the student researcher seek out other works detailing college and university
television and cable operations. Such works as Silvia and Kaplan's *Student
Television in America: Channels of Change* (1998) help to provide insight into
these operations showing the complementary aspects of college radio.

This book is also intended to complement current pieces on radio on the college campus, including material continually provided by such organizations as
the National Association of College Broadcasters (NACB), the Intercollegiate

Broadcasting Service (IBS), and the College Media Advisors (CMA). Additionally, the materials and support provided by sources such as the Broadcast Education Association (BEA), the National Broadcasting Society (NBS/AERho), and the Association for Education in Journalism and Mass Communication (AEJMC) are also noted.

The programming provided by college broadcasting is normally grouped into three areas: entertainment, education, and information. As you read this book, you will also become aware that college broadcasting provides the very important academic role of training students for careers in broadcasting. To this end you will find that many of the ideas, applications, and trends described throughout the book are also applicable to college television and cable operations (see Sauls 1996). And, of course, probably the entire concept of college radio can be applied to high school radio. Yes, some high schools and school districts operate radio stations (see Haber 1996).

Classroom Use

Again, the practical applications, along with the general overview, will be useful in the classroom function of the book to address telecommunications in higher education. Thus, for the student this book serves to introduce, explain, and explore the world of college radio broadcasting in America.

Academically, the manuscript was prepared as a library resource. Here, the presentation of the material takes a form consistent with a typical research style (citations, quotes, references, and so on). Upper-level undergraduates and graduate students should be able to rely on the book for their research on college radio and overall radio broadcasting. The author encourages the researcher to seek out the references made within the book (no matter how dated) to gain even further insight into the subject.

In the classroom, more applicable textbooks are available to provide overviews of radio and broadcasting/cablecasting in general, such as Head et al.'s *Broadcasting in America: A Survey of Electronic Media* (1998), and Smith, Wright, and Ostroff's *Perspectives on Radio and Television: Telecommunications in the United States* (1998). Keith's *The Radio Station* (1997) and Adams and Massey's *Introduction to Radio: Production and Programming* (1995) are also appropriate for use. Any course, though, emphasizing college radio will find the text in hand most applicable. For instance, I teach a university graduate-level course on Public Telecommunications Management, which highlights college and university broadcasting, including radio. In this venue this book could complement such textbook resources as Sherman's *Telecommunications Management: Broadcasting/Cable and the New Technologies* (1995). If anything, I hope that this book will spawn classroom discussion concerning all aspects of college broadcasting. Additionally, I recommend that graduate students who are venturing into the world of survey research and college radio stations considering their own listener research (see Chapter 5) consult such academic works as Alreck and Settle's *The Survey Research Handbook* (1985), as well as outside expertise.

It is intended, then, that this book provide an overview to the reader about what college radio is, why it exists, and how it is presented in the United States. Of particular interest are the implications imposed by college radio on the contemporary media scene, and the role such a medium plays in American culture.

REFERENCES

Adams, M. H., and K. K. Massey. (1995). *Introduction to radio: Production and programming.* Madison, Wis.: Wm. C. Brown Communications.

Alreck, P. L. and R. B. Settle. (1985). *The survey research handbook.* Homewood, Ill.: Richard D. Irwin.

Brant, B. G. (1981). *The college radio handbook.* Blue Ridge Summit, Pa.: TAB.

Gimarc, G. (1994). *Punk diary: 1970-1979.* New York: St. Martin's Press.

———. (1997). *Post punk diary: 1980-1982.* New York: St. Martin's Press.

Haber, A. (1996). High schools breed new crop of radio stars. *Radio World* 20(14): 6.

Head, S. W., C. H. Sterling, L. B. Schofield, T. Spann, and M. A. McGregor. (1998). *Broadcasting in America: A survey of electronic media,* 8th ed. Boston: Houghton Mifflin.

Keith, M. C. (1997). *The radio station,* 4th ed. Boston: Focal Press.

McCluskey, J. J. (1998). *Advising, managing and operating a successful student/noncommercial radio station.* Needham Heights, Mass.: Simon and Schuster.

———. (1998). *Starting a student/noncommercial radio station.* Needham Heights, Mass.: Simon and Schuster.

Sauls, S. J. (1996). College radio: Points of contention and harmony from the management perspective. *Feedback* 37(2): 20-22.

Sherman, B. L. (1995). *Telecommunications management: Broadcasting/cable and the new technologies,* 2nd ed. New York: McGraw-Hill.

Silvia, T., and N. Kaplan. (1998). *Student television in America: Channels of change.* Ames: Iowa State University Press.

Smith, F. L., J. W. Wright II, and D. H. Ostroff. (1998). *Perspectives on radio and television: Telecommunications in the United States,* 4th ed. Mahwah, N.J.: Lawrence Erlbaum.

2

College Radio Stations: What Are They?

Culture is a historically rooted, socially transmitted set of deep patterns of thinking and ways of acting that give meaning to human experience, that unconsciously dictate how experience is seen, assessed and acted on. Culture is a concept that helps us perceive and understand the complex forces that work below the surface and are in the air of human groups and organizations.

—T. E. Deal and K. D. Peterson, *The Principal's Role in Shaping School Culture*

The term "college radio" encompasses stations operating on college and university campuses, including two-year colleges. Initially, college radio stations were developed as experimental stations. Today college radio broadcasting comes in many forms, in addition to the standard FM and AM radio station. Carrier-current/wired-wireless and closed-circuit stations offer campuses additional outlets. An even more recent offering is the FM cable-access station provided to schools through local cable-TV systems or on-campus networks (Sauls 1995, 1-2). College and university radio stations are operated in an ongoing manner. Basically, these stations are run just like a business or an "auxiliary enterprise" on campus.

THE BEGINNING OF EDUCATIONAL RADIO STATIONS

When discussing the background of educational radio, one can present a great deal of history. Here, though, only a brief overview is provided to introduce the reader to college radio, serving simply to illustrate how we arrived at the place we are today in this medium. This section and the next, along with the historical perspective presented at the end of this chapter, are included to establish the context in which college radio exists.

Throughout my initial discussion of college radio, particularly in relation to its being a "telecommunications medium" in general, reference is made to a 1977 work entitled *Educational Telecommunications* by Donald N. Wood and Donald G. Wylie. If we wish to learn about the true "pioneers" of college radio, all early educational radio should be considered. From the beginning, Wood and Wylie note that in 1917, three years before KDKA (Pittsburgh) and WWJ (Detroit) were to start their "conflicting" claims as the first radio station on the air, in Madison, the University of Wisconsin started experimental broadcasts over its radio station, 9XM (later to adopt the call letters WHA). They also note the first educational institution officially licensed to operate a radio station was the Latter-day Saints' University of Salt Lake City, Utah, "on an unlisted day and month in 1921" (Frost 1937, 178). Other early college radio stations authorized to operate included the University of Wisconsin and the University of Minnesota, which received their official broadcast licenses on January 13, 1922 (Frost 1937, 464).

Thus these authors establish (and, in sum, give credit to) the very first college radio stations—a debate that goes on today (see Avery 1998). Some credit the Brown University radio station in Providence, Rhode Island, as being the first "college radio station."

The authors continue their discussion concerning the beginning of educational broadcasting, noting that most of the stations remained on the air less than three years. These "early failures" can be attributed (as noted by the authors) to several factors, among which were insurmountable financial problems, novelty engineering experiments, and formidable commercial opposition, including disputes over broadcast channel assignments, whose profit value was rapidly becoming apparent.

The authors note that the educators themselves were slow to see the real possibilities of the new medium. "Regardless of the explanations offered, it is clear that educators were generally apathetic toward educational broadcasting" (Saettler 1968, 205). "They failed to see its educational potential; they failed to grasp its public relations value" (Wood and Wylie 1977, 20). As one educational broadcasting pioneer stated more than 20 years later (quoted by the authors): "The stations were peripheral to the main business of the institution" (Hull 1956, 7).

EDUCATIONAL RADIO CHANNELS

"The FRC [Federal Radio Commission, established by Congress in 1927] frequently restricted educational stations to low power, gave them poor frequencies, or in many cases where time was shared with a commercial station, assigned the least desirable operating time to the educational station" (Hill 1954, 11). Wood and Wylie (1977, 30) note that the first recorded official request for reserved educational broadcasting channels came out of the 1930 convention of the ACUBS (Association of College and University Broadcasting Stations). (The aspect of "educational radio" today is addressed in Chapter 4 within the venue of alternative programming.)

Beginning in 1944, the Federal Communications Commission (FCC), established under the Communications Act of 1934, held hearings that resulted in the reservation of a total of twenty FM channels for educational use. "These channels, located between 88.1 and 91.9 MHz inclusively, remain the section of the FM spectrum reserved for noncommercial radio" (Wood and Wylie 1977, 32; see also Runyon 1998). Little did the FCC know then that these channels would be so precious a commodity by the 1980s!

Thus the growth of noncommercial educational FM radio, generally the staple of college and university radio stations, can be attributed to the Federal Communications Commission's allocation on June 27, 1945, of twenty FM channels set aside exclusively for noncommercial use (channels 201 to 220 between 88 and 92 megahertz, Federal Communications Commission 1946, 20-21; see Fig. 2.1). It is interesting to note that prior to the 1945 allocation, channels for educational use were established as "so-called curricular channels

SUBPART C

NONCOMMERCIAL EDUCATIONAL

FM BROADCAST STATIONS

73.501   CHANNELS AVAILABLE FOR ASSIGNMENT.

(a)   The following frequencies, except as provided in paragraph (b) of this section, are available for noncommercial educational FM broadcasting:

Frequency (MHz)	Channel No.	Frequency (MHz)	Channel No.
87.9	200	90.1	211
88.1	201	90.3	212
88.3	202	90.5	213
88.5	203	90.7	214
88.7	204	90.9	215
88.9	205	91.1	216
89.1	206	91.3	217
89.3	207	91.5	218
89.5	208	91.7	219
89.7	209	91.9	220
89.9	210		

(Note: The frequency 87.9 MHz, Channel 200, is available only for use of existing Class D stations required to change frequency.)

FIGURE 2.1 Noncommercial Educational Channels
Source: Part 73, *Radio Broadcast Services,* Rules Service Company, 1994–95, Record 1361/4419. Used with permission.

by the FCC in 1938" (Avery 1998, 83). The number of noncommercial educational FM licenses increased from 38 in June 1947 (Avery and Pepper 1979, 22) to more than 1,100 college, university, and school-owned radio stations in 1997 (*Broadcasting and Cable Yearbook 1997,* B578-B579). Overall, the number of public, noncommercial radio stations increased from 396 in 1969 to 1,076 in 1980 (Public broadcasting 1981, 79). This rapid increase in the number of noncommercial educational FM radio stations is attributable to the recognition by colleges and universities of their potential as academic training facilities, community service outlets, and, most important, public relation arms for the colleges and universities. "By the mid-1990s, more than half of the 1,800 noncommercial educational 'public' radio licences were still held by institutions of higher learning" (Avery 1998, 83). Concerning noncommercial educational broadcast services, the Federal Communications Commission stated as early as 1948 that "[s]tations in this service are used principally by universities and school systems for transmitting educational and entertainment programs to schools and to the public. Their operation is entirely on a noncommercial basis" (Federal Communications Commission 1948, 44).

This is why almost all broadcast college radio stations are located to the far left of the FM dial—they are noncommercial entities. One difference in listening to stations in the noncommercial educational portion of the FM band is that the "channel separation" is not as defined as in the commercial portion above 91.9 megahertz. This is because noncommercial stations are not required to have the same minimum cochannel separation as in the commercial portion of the FM band and are actually allowed an interfering contour along with its protected contour. For the station a consulting engineer can ensure that proper separation is being provided within legal parameters.

But, as described in Chapter 7, it may not be a far-fetched idea for a college or university to consider a commercial license. The fact of the matter is that such an enterprise would allow for the station to seek true potential advertisers. A college radio station does not have to be a noncommercial station. Here, then, the aspect of running a business (the college radio station) truly comes into play. This aspect is addressed next in this chapter under the ideals of operating auxiliary enterprises on campus.

Also, as discussed later in this chapter, the college radio station does not have to be a "broadcasting" entity. It is possible for stations to broadcast only on campus via wire/wireless, radiating cable, or closed-circuit. Additionally, it is possible for the station to broadcast to the community at large via a cable channel. In fact, this has become very popular in the last 10 to 15 years, as it enables college stations to "get on the air" without encumbering the huge costs associated with a broadcasting infrastructure (transmitters, antennas, and towers) and undergoing the process (which can be long and expensive) of securing a broadcasting license from the FCC. Then there's webcasting/broadcasting on the Internet.

Each individual college and university has its particular wants and desires to meet its needs. Therefore, the types of college radio station designs (as described in Chapter 3) vary from school to school. As radio frequencies are becoming less and less available, schools should consider numerous avenues for radio opportunities. Furthermore, the importance of operating the non-broadcast station as if it were an "over-the-air" broadcast facility should be emphasized to the students. Often the students will not take as seriously as they would a licensed station the operation of a cable-only outlet. This philosophy needs to be presented via the station advisor (see Chapter 6), including the actual program content presented.

AUXILIARY ENTERPRISES

Basically, college and university radio stations are run just like a business or "auxiliary enterprise" on campus in the sense that they are ongoing "service operations conducted to the benefit of students and faculty" (Ohio House of Representatives 1969, 57). Furthermore, these stations exist to provide a service to the community through their broadcasts.

However, in contrast to typical auxiliary enterprises operated on college or university campuses, noncommercial radio stations normally do not charge a "fee directly related to, although not necessarily equal to, the costs of the goods or services" (Hughes 1980, 96). The campus radio station is typically funded directly by the school, either through direct funding or some type of student service fee allocation (see Chapter 7 on funding). As an auxiliary enterprise, then, the noncommercial college and university station is completely controlled and funded by the individual institution, and thus the station's continuance, "expansion or curtailment does not require state approval nor are state funds made available for these purposes" (Ohio House of Representatives 1969, 57). However, even under such funding restraints they are expected to function and operate in an ongoing manner, emulating their commercial counterparts. Thus, in reality, the campus radio station is a business operation and designated as such. And though it is expected to function daily, it usually does so with reduced funding, staffing, and support when compared to most other entities on campus (see Sauls 1996, 20).

The National Association of College and University Business Officers sponsored Welzenbach's 1982 book entitled *College and University Business Administration.* One of the six parts in the authoritative work addresses business management and the area of auxiliary enterprises. According to Welzenbach, "the distinguishing characteristic of most auxiliary enterprises is that they are managed essentially as self-supporting activities, although sometimes a portion of student fees or other support is allocated to assist these activities" (198). This mirrors the typical college radio station alignment. College radio stations, as auxiliary enterprises, "are recognized vehicles for attracting and retaining students, faculty, and staff" (198).

In 1987 Barnett pointed out that "a vital and dynamic force has emerged on the college campus—the Auxiliary Operations!" (31). In an article published in *College Services Administration,* the journal of the National Association of College Auxiliary Services, Barnett included a complete list of 80 auxiliary areas, including radio stations, that function on campuses in the United States and Canada. Although he concluded "it is an exciting explosive period for the auxiliaries and a challenging one for the directors who manage this vital aspect of college and university life" (32), direct funding of specific units, particularly radio stations, was not addressed.

In a 1990 article, Noetzel and Hyatt cited Barnett's recognition in 1987 of radio stations as potential auxiliary services (129). In a thorough discussion of auxiliary enterprises Noetzel and Hyatt stated that "sound financial management of the auxiliary areas of colleges and universities is essential to the fiscal health of the institution" (117). They also noted that "the quality and level of services provided by auxiliary operations ... can be an effective incentive for recruiting and retaining students" (118). Topics such as soliciting student input, confronting budget reductions, modern management techniques, and major issues facing auxiliary units on campuses were also discussed by Noetzel and Hyatt (see Chapter 2 in Sauls 1993). All of these issues are addressed continually in the operation of the college and university radio station.

TV, CABLE, AND COLLEGE RADIO

Although management styles and theories differ among administrators of college radio stations, views concerning the operational aspects of college radio, both on and off campus, tend to be consistent. As noted in Chapter 1, "generalizations" addressing the operation and management of college radio stations can also be applied to college-operated television and cable outlets. Common factors are those tendencies inherent to the administration of all college and university broadcast/cable media (see Sauls 1996, 20).

With the utilization of cable by radio (billboard/scroll and audio services), it is imperative that personnel be aware of all operating-entity characteristics. In other words, become familiar with the operation and programming of radio, television, and cable services. This is where the "mass communications" education and training pays off!

In addition, as stated earlier in this chapter, the use of cable television (encouraged by necessity due to the availability of fewer broadcast frequencies) provides an avenue to radio utilization. This, of course, is in lieu of (or in addition to) over-the-air broadcasting. The college radio station may be the ongoing audio for a video scroll provided by the school, either 24 hours per day or when the cable television station is not providing programming. Cable may in fact be the only choice for any radio "broadcasting." This type of set-up allows for great cross-utilization of both radio and television students.

PRACTICAL APPLICATIONS

Making Contacts

A good starting point is to see what everyone else is doing. Many listservs on the Internet and home pages on the World Wide Web provide the opportunity to ask questions and discuss options. A few of the lists and sites concerning college radio are:

Airwaves Radio Journal (Also allows for routing through *rec.radio.
    broadcasting.*)
  Moderator's Mailbox: *MODERATOR@airwaves.com*
  Submit Articles To: *ARTICLES@airwaves.com*
  World Wide Web: *http://www.airwaves.com/*
  Archives: *http://www.airwaves.com/AIRWAVES_ONLINE*
  SUBSCRIBE@airwaves.com

Broadcast Education Association
  http://www.beaweb.org

College Media Advisers
  http://www.collegemedia.org/

College Radio DJ Discussion List
  DJ-L@listserv.nodak.edu
  ListServ@ListServ.syr.edu

Intercollegiate Broadcasting System
  http://www.ibsradio.org./ibshome.html

National Association of College Broadcasters
  NACB@lLISTSERV.SYR.EDU
  http://www.hofstra.edu/~nacb/

National Association of College Broadcasters (Faculty)
  NACB-FS@lLISTSERV.SYR.EDU

National Broadcasting Society/Alpha Epsilon Rho (AERho)
  http://www.onu.edu/org/nbs

Play By Play Discussion List
  PBP-L@listserv.syr.edu

Public Radio Discussion Group
  PUBRADIO@LISTSERV.IDBSU.EDU

The NACB, IBS, NBS, NFCB, CMA, and BEA

Another practical way to learn about college radio is to join organizations that promote student broadcasting, or at least to attend their conferences. Here numerous groups are addressed. Descriptions of the various organizations are presented only as a means of informing the reader and do not indicate preferences of one group over another. Each individual station will have specific characteristics and needs that certain groups can more adequately meet. These descriptions are provided as an introduction to various sources.

Throughout this book, reference is made to an organization known as The National Association of College Broadcasters (NACB). According to the *1995 NACB Station Handbook,* the purpose of the NACB is:

> To provide an exchange of ideas, programming and information in the student media community and among students, faculty and professionals; To facilitate growth, prestige and recognition of the student media community; To protect and lobby for the laws and regulations that affect the student media community to the extent allowed by law; To encourage and assist student stations and individuals in attaining high standards which will enhance the communities served; To provide opportunities for individuals with an interest in media and communications; To support student media endeavors and encourage unique and creative innovations in that community. (220)

Prior to 1998 the NACB was highly recommended for college radio, television, and cable station membership. If nothing else, the NACB provided excellent advice on running a station. Unfortunately, as of 1999 the NACB was dissolved. A new group, Collegiate Broadcasters, Inc. (CBI), has been formed.

Another group dedicated to college radio broadcasting is the Intercollegiate Broadcasting System. See Figure 2.2 for information regarding the IBS.

The National Broadcasting Society (NBS) was formerly known as Alpha Epsilon Rho (AERho) when it was the National Honorary Broadcasting Society. It has been in operation for more than 50 years with literally thousands of members during its existence. The NBS offers regional and national conventions, relying heavily upon local chapters for yearly activities on campuses and within designated regions. The author can attest to NBS's scholarship efforts in providing much-needed financial support to deserving students. As of 1999 AERho had formed an alliance with the International Radio and Television Society to create an honor society within the IRTS (*http://www.onu.edu/org/irts-aerho/his.hmtl*).

About IBS...

- **Who is IBS - what does it do...,
  and how can IBS membership benefit your station?**
- **How can your station join IBS?**
- **How do you get in touch with IBS?**
- **How can you start a new station at your school or college?**
- **How can you increase your station's coverage?**

Who is IBS...how can it benefit your station?

IBS is a nonprofit association of student-staffed radio stations based at schools and colleges across the country. Some 600 member-stations operate all sizes and types of facilities including closed-circuit, AM carrier-current, cable radio and FCC-licensed FM and AM stations.

IBS was founded in 1940 by the originators of AM carrier-current campus radio. Originally, most of the interest involved exchanging technical information among colleges on this new form of transmission. As more stations became established and grew, the interest evolved to include management, programming, funding, recruiting, training, and other operational and creative areas.

Since AM carrier-current stations are permitted to carry commercials, a strong emphasis was initially placed on commercial advertising. Remember - at that time, FM broadcasting was not a factor so AM carrier-current stations were able to appeal more directly to those who had been listening to local commercial AM stations.

When the FM broadcast band was shifted to its present band (88.1 - 107.9 MHz), IBS people were active in efforts to reserve a group of frequencies specifically for noncommercial educational use. The result was the creation of the reserved band (88.1 - 91.9 MHz) where most noncommercial stations are now located. IBS was also active in convincing the FCC to establish the category of Class D (10-watt) noncommercial FM stations as an entry level. This encouraged hundreds of new stations to get started, most of them eventually increasing power to at least minimum Class A facilities of 100-watts.

In more recent years, when the FCC decided that bigger was better, (at least for noncommercial FM), IBS was the only national organization actively fighting for retention of 10-watt FM stations.

When restrictive rules threatened the future of carrier current stations, IBS, working with equipment vendors, got the FCC to relax the rules and recognize the importance of these stations.

FIGURE 2.2 IBS (Intercollegiate Broadcasting System)

Source: Intercollegiate Broadcasting System, Inc. Used with permission.

Two other organizations warrant mentioning: the NFCB and the CMA. The National Federation of Community Broadcasters' (NFCB) annual conference includes workshops on such topics as digital production, listener habits, and donor methodology for community stations. (See Annual community radio conference 1996.) In the past the NFCB also distributed an in-depth legal handbook for public radio.

The College Media Advisers (CMA), in conjunction with the Association Collegiate Press, holds an annual National College Media Convention. As of 1998 the CMA was expanding its broadcast panels at its convention to assist in picking up where the NACB left off. The CMA convention is also very journalistic oriented, with more than 2,200 attendees at past conferences (see Fig. 2.3).

Finally, I would be remiss if I didn't mention the Broadcast Education Association. With some 1,400 individual and over 225 institutional members, the BEA is an "organization for professors, students and professionals involved in teaching and research related to radio, television and electronic media education" (*http://www.beaweb.org/main.html*). The BEA publishes quarterly the *Journal of Broadcasting and Electronic Media* and *Feedback,* and biannually the *Journal of Radio Studies.* Additionally, the BEA distributes a comprehensive annual *Membership Directory* (of which I am the editor at the time of this writing). The BEA's annual convention is held in conjunction with that of the National Association of Broadcasters.

Station Characteristics

To provide a general view of college and university radio stations, the following attributes found in The National Association of College Broadcasters' *1995 College Radio Survey* (iv) are provided. (I provided the survey design and data analysis coordination for the study.) These factors furnish a "snapshot" view of the medium known as "college radio." (More information on these points is available in the Survey Findings section of the *College Radio Survey.*)

■ Seventy percent of the stations were FCC licensed, and 6.1 percent were carrier current. Closed-circuit, radiating-cable, NPR, and cable-television stations each made up less than 4 percent of the sample. Stations using several combinations of broadcast methods were also represented.

■ Fifty-six percent of the broadcast stations operated in the 100 to 3,000 watt range, and 29.8 percent operated above 3,000 watts.

■ The majority of the respondents from cable-television stations reported that they are received by fewer than 10,000 households.

■ The majority of the respondents from carrier-current, closed-circuit and radiating-cable stations reported that they had 100 percent access to rooms of students in residence. Other areas in which these stations were made available to students included cafeterias/dining areas, student unions/student centers, and lounges/recreation areas.

What is College Media Advisers?

CMA front page

Updated 06.09.98

Join CMA.
Join our e-mail
discussion list.

**Questions about
CMA?**
Send e-mail to
Ron Spielberger,
CMA executive director.

Questions on the site?
Send e-mail to
Ron Johnson,
CMA site coordinator.

CMA represents the people who advise the nation's collegiate newspapers, yearbooks, magazines and electronic media.

With more than 700 members from coast to coast, College Media Advisers has been working since 1954 to support both new and veteran advisers of college-media programs.

CMA serves thousands of students and advisers at two annual conventions. The CMA Newsletter informs members of trends and news, and College Media Review, our flagship journal, is the leading academic journal on advising collegiate media, both print and electronic.

CMA mission statement

As the professional association dedicated to serving the needs of collegiate student media programs and their advisers, our mission is to --

● educate and inform advisers about their roles in serving students and abou the teaching, advising and production of collegiate media.
● advance the aesthetics of the student media our members advise and the technologies of these programs.

To achieve this mission, we will --

● offer conventions, workshops, conferences, seminars and similar learning experiences.
● serve as the authoritative voice of the collegiate media and advisers,
● provide and disseminate research and information for and about collegiat media and advising.
● provide a learning environment that promotes excellence and ethical standards in all aspects of mass communication.

Approved March 17, 1993

FIGURE 2.3 CMA (What is College Media Advisers?)
Source: College Media Advisers. Used with permission.

■ Eighty-eight percent of the respondents reported that their stations operate at least 12 hours a day. This includes 34.2 percent that operate 24 hours a day, 25.7 percent that operate 18 to 23 hours a day, and 28.4 percent that operate 12 to 18 hours a day.

■ Fifty-one percent of the respondents reported that their stations operate 365 days a year. Twenty-six percent reported that they operate fall and spring semesters only, and 23.4 percent reported that they operate either one semester only or summer only.

Starting a Student Radio Station

The following section from the National Association of College Broadcasters' *1995 NACB Station Handbook* (13-14) provides an excellent description of the types of stations available for colleges and universities to consider for use:

> Be aware that building a station from idea to reality is a long process—on average, 2-3 years or more. Thus if the primary organizers are upper-classmen, enlist support from an encouraging faculty member and/or younger pro-station students from the start. Do not ignore school Administration (and the communications department, if there is one). The supervision and help from a faculty member or administrator who supports the effort to start a college radio station is essential; the chances of success are far greater than if students alone undertake the effort. The long time frame involved and the political connections with university entities essential are two key reasons for taking on a faculty advisor when beginning a station start-up project. Technically speaking, college radio stations come in one of four types: carrier current, cable FM, low-power (experimental) broadcast or full-power (regular) broadcast. (To date, digital audio broadcasting has not yet been implemented in the United States.)
>
> *Goodday from _____-FM in _____. I would like to post a question concerning low-wattage AM broadcasting. Our station is looking into the addition of a training wing that will also better serve the community with more news and sports coverage.*
>
> *What do we need to consider when looking into this proposal? We plan to cover the campus of the University of _____, which, for our purposes, is about a five-mile radius. How do we select frequency and call letters? If anyone has been in a similar situation, please respond. (NACB Discussion List, July 24, 1996)*

Low-Power Broadcast

> A low-power broadcast FM [or AM] station may be the most cost-effective solution if the goal is to reach a limited geographic area without the limitations of wires. Note that the FCC-mandated minimum station power output is [100 watts for noncommercial FM and 250 watts for noncommercial AM] for [licensed] stations. As an alternative to carrier current AM, the FCC allows for analogous low-power FM transmitters (but still with a very limited broadcast radius). (National Association of College Broadcasters 1995a, 14)

Information regarding low-power broadcasting can be obtained from the Federal Communications Commission Web site at: *http://www.fcc.gov/mmb/asd/lowpwr.html*. Also of note, as of 1998, the FCC was considering proposals to create a new power broadcasting service (see Cole 1998). On a historical note, in 1948 "[r]ules to permit low-powered educational FM broadcasting became effective September 27 and, on October 21, the Commission granted the first construction permit for a noncommercial educational station with power of less than 10 watts, to Syracuse University" (Federal Communications Commission 1950, 6).

Full-Power Broadcast

> Over-the-air college stations typically reside on the FM band between 88.1 and 91.9. This spectrum of the radio dial is reserved for non-commercial, educational stations, of which college radio is one type. Though the chances of a college getting a station here are better, in many markets this part of the band is just as crowded as the commercial band. And since a commercial licensee allows advertising, the station could pay for itself. The broadcast license to operate a station on a given frequency comes from the FCC, and is granted only after approval of an application. The most common application pursued by a fledging college station would be Class A station (the lowest category), which requires a minimum 100 watts of station output. (National Association of College Broadcasters 1995a, 14)

Furthermore, you will find college stations operating "regular" signals on the AM band, also known as the "Standard Broadcast Band"—a consideration if the FM band is full in your market.

> Significant fees are involved in applying for a broadcast station construction permit (CP), both to the FCC for processing and to the consulting engineer who must be hired to: (1) do the required frequency search (to determine if an open frequency exists in the market where a new station could go without causing interference to other existing stations); and (2) conduct the transmitter/ tower site survey and other requirements to complete the technical portions of the FCC application. If there are no problems, the application approval process takes 6-12 months on average. Add station construction time and it could be two years by the time the station is on the air. Hiring a communications lawyer (Washington, D.C., is full of such firms) may also be worthwhile to avoid problems later. (National Association of College Broadcasters 1995a, 14)

As mentioned earlier in the chapter, a station can also broadcast to the community at large via a cable channel. Here the station is the audio portion of the TV program, quite often an information "scroll." While it is not "broadcasting over the air," at least the station is being "transmitted" in a way via cable. Of course, the limitation rests in the fact that the station will be received only by those with cable and most likely be limited to local cable companies. But it's better than nothing, and you don't have to deal with any FCC regulations.

Limited Area Broadcasting and Unlicensed Operations

"Limited Area Broadcasting" falls under Part 15 of the FCC Rules and Regulations as applicable for educational and noncommercial stations. (Information regarding Part 15 devices may be obtained via the Federal Communications Commission Web site at *http://www.fcc.gov/mmb/asd/ main/other.html*.) Note that these systems, properly designed, are for use to provide a broadcast signal to either an entire campus or portions of a campus.

Here, then, a school may wish to look at an unlicensed station. John E. Devecka, sales manager for LPB, Inc., details in Figure 2.4 options for such operations. More information may be obtained from the LPB Web page at *http://www.lpbinc.com.*

Many schools operate more than one radio outlet. Any combination of signals allows for greater student participation, more diversity in programming, and increased community service. Additionally, a successful commercial station on campus could help support its noncommercial counterpart(s). See Chapter 7 for an example of multiple ownership at the University of Florida.

This also raises the subject of "broadcasting" over the Internet, both under the titles of webcasting and netcasting. Details of utilization of these mediums are discussed in chapters 7 and 9. The point to be made here is that there is no FCC jurisdiction in regard to broadcasting on the Internet, as such a broadcast is not going out "over the airwaves."

Concerning equipment, the *1995 NACB Station Handbook* states:

> Expense is obviously a concern. Regardless of the type of station to be constructed, a broadcast studio will be needed. Equipment for programming and production is expensive. Used equipment is cheaper, but usually will not perform as well as new pieces, even if refurbished. Try to buy new for the key components of the station, even if tight budgets force the purchase of used units for the rest of the equipment. Obtaining free equipment from area commercial stations is possible, since donations to colleges are tax write-offs. Commercial stations up-grade periodically; contact them through the station engineer. However, if the donated pieces are over a few years old, breakdowns and repair costs might outweigh asking prices on new models. (14)

If the school decides to pursue a broadcast station, then formal application must be made with the Federal Communications Commission for a station license as prescribed by the Communications Act of 1934 (and as amended by the Communications Act of 1996).

Nonlicensed broadcast operations are known as "pirate radio." The U.S. District Court for the Northern District of California ruled in June 1998 upholding FCC radio licensing authority by issuing an injunction against a pirate broadcaster. The decision by the court affirmed "the FCC's authority to require a license before any person can broadcast on the public airwaves" (FCC News, June 17, 1998, Report No. G 98-10, *http://www.fcc.gov.cib/News _Releaseas/nrci8011.html*). This ruling also upheld the constitutionality of the Federal Communications Commission's broadcast licensing procedures. Rulings against pirate radio operators can result in penalties including fines, seizure of radio equipment, and imprisonment (see Texas Association of Broadcasters 1998, 10).

Of course, the basis of the pirate radio argument is freedom of speech ver-

So, What Kind Of Unlicensed Station Can I Start?

AM Carrier Current

- Systems place AM signal on electrical lines of the campus
- Mono Signal
- Anywhere on AM band 530-1700kHz*
- No FCC License needed
- Typically one package (Transmitter + Coupling Unit) per building
- Commercials permitted
- Typically maximum of $1150/building, generally much less
- Site Survey Recommended

AM Vertical Antenna

- Systems broadcast via vertical antenna, typically 20ft in height
- Mono Signal
- Anywhere on AM band 530-1710kHz*
- No FCC License needed
- Typically one package per campus, outside coverage only
- Commercials permitted
- Typically maximum of $2500/campus
- Site Survey Recommended, Careful Engineering REQUIRED

FM Radiating Cable

- Systems broadcast via radiating coaxial cable, installed inside buildings on one floor
- Mono/Stereo Signal
- Anywhere on FM band 88-108MHz*
- No FCC License needed
- Typically one package (Transmitter + Radiating Cable) per building
- Commercials permitted
- Typically maximum of $2700/building, generally much less
- Site Survey Recommended, Post-Installation Testing REQUIRED

FM Cable TV Broadcasting

- Systems place FM signals on campus or local Cable TV systems
- Mono/Stereo Signal
- Anywhere on FM band 88-108MHz*
- No FCC License needed
- Typically one package (Transmitter) per campus
- Commercials permitted, subject to approval by CATV operator
- Typically maximum of $1110/campus
- Site Survey Not Needed

* FCC Part 15 requires that these stations may not cause interference to any licensed operation.

LPB offers these systems as well as commercial and non-commercial licensed station solutions.

 28 Bacton Hill Rd., Frazer, PA 19355. P: 610-644-1123, F: 610-644-8651, http://www.lpbinc.com

FIGURE 2.4 LPB, Inc. ("What kind of nonlicensed station can I start?")
　　　　Source: LPB, Inc. Used with permission.

sus illegal interference with licensed broadcasters. Michale H. Bader presents an in-depth description of the pirate radio dilemma in a 1998 article in *Radio Ink* magazine entitled "Anarchy on the Airwaves: The Growing Threat of Pirate Radio." Bader provides the address of the Free Radio Network at *http://www.frn.net/* to acquire more information about "a variety of pirate topics" (61).

In general, the following steps outline the procedure for broadcast licensure:

1. An Application for Construction Permit is filed for either (a) Commercial Broadcast Station or (b) Noncommercial Educational Broadcast Station (the most common for college and university radio stations, as discussed earlier). If you are acquiring an existing station, then an application for Consent to Assignment of Broadcast Construction Permit or License is filed. Radio station brokers and communications attorneys are good sources in locating existing stations for sale or transfer.
2. Local Public Notice is run by the applicant in a newspaper of general circulation in the community where the station is or will be located.
3. A 30-day filing period is designated in which competing applicants may file to oppose or petition to deny any application.
4. Applications are reviewed by the FCC Mass Media Bureau in regard to engineering, legal, and financial data.
5. If needed, the FCC may designate the application for hearing in regard to application conformity and/or in response to oppositions and petitions to deny (no. 3).

   Here it is most important that the applicant be aware of current rules in regard to what is known as "competitive hearings" for competing applicants. Quite often rules governing noncommercial stations are set apart from those for commercial stations. Again, this is where keeping up-to-date on the current rules and regulations is extremely important. For example, on October 21, 1998, the FCC was seeking comments regarding the use of either lotteries or a point system to choose between competing applications for noncommercial broadcast stations as opposed to comparative hearings (Federal Communications Commission 1998: FCC 98-269).
6. A construction permit is issued specifying the period time in which the construction of the station is to be completed.
7. Upon completion of program testing, a station license is granted (eight years for radio).

Further specifics regarding station application and licensing procedures can be obtained from the Mass Media Bureau of the FCC.

As of 1998 the Federal Communications Commission was pursuing the concept of auctioning spectrum allocations. The rules released in August 1998 pertaining to auctions addressed only commercial licenses. "The FCC invited further comments on whether noncommercial applicants competing for licenses on the commercial FM band should participate in auctions as well. That ongoing proceeding is MM Docket 95-31" (Stimson 1998, 14). Of course, any action pertaining to any noncommercial station can impact current and future operations (see chapters 8 and 9).

As alluded to earlier, be aware that the licensing process can take a long time. For example, my own experience provided an eight-year process from original application filing to commencement of programming for a power increase at the university station I managed. Included was attendance at an administrative hearing in Washington, D.C., as noted in number 5 of the preceding list. In the next chapter, I highlight the need for stations to consider acquiring both outside legal and outside engineering advice when warranted.

For legal matters, sources include Rules Service Company, *Broadcasting and the Law,* and Pike and Fischer's FCC subscription news service. Additionally, two online-accessible government sources are worth noting for station operational matters. The FCC Web-page address is *http://www.fcc.gov/mmb/asd/.* This address will take you directly to the Audio Services Division site.

> The FCC itself does not keep a public database of its rules sections. That task
> is performed by the *Government Printing Office* for a large number of agencies.
> These rules and regulations are compiled in the *Code of Federal Regulations
> (CFR). ...* Telecommunications falls under Title 47 of the CFR. AM, FM, and
> TV broadcast stations fall under Part 73 of Title 47. Additional rules sections
> pertaining to radio broadcasting are contained in Parts 0 and 1 of Title 47.
> (Radio Broadcasting Rules—47 CFR Part 73 [FCC] USA. *http://
> www.fcc.gov/mmb/asd/bickel/amfmrule.html*)

The Code of Federal Regulations Web-site address is *http://www.access. gpo.gov/nara/crf/index.html.*

To locate existing stations in your area, the following resources are available: a database of all college and university (and some high school) radio stations from the National Association of College Broadcasters, the *M Street Radio Directory,* the *FM Radio Atlas,* and the *Broadcasting and Cable Yearbook.* (As always, stations, as nonprofit entities, should request an educational discount on [or even gratis] subscriptions.) Lists of stations that have recently gone off the air and thus are considered "dark" by the FCC can be located at *http://www.fcc.gov/mmb/asd/* (then add either *amsilent.html* for AM Standard Broadcast stations, or *fmsilent.html* for FM stations).

A point to be addressed, as raised in Chapter 1, concerns high school stations. With the growing number of stations licensed to high schools and school districts, many of tomorrow's college broadcasters are gaining true experience at the high school level. Additionally, such events as the Marconi College Radio Awards held at the annual Loyola Radio Conference in Chicago, acknowledging excellence in both college and high school radio, is an example fostering "precollege" radio. For example, school stations in the greater Boston area include WBMT-FM (Boxford), WBPV-FM (Charlton), WHHB-FM (Holliston), WWTA-FM (Marion), WRPS-FM (Rockland), WSDH-FM (Sandwich), and WSRB-FM (Walpole) (*Broadcasting and Cable Yearbook 1997,* B210-B214).

Note that the detail of starting a station does not address such factors as programming (music, for example) or status (noncommercial or commercial). These aspects are brought forth in the following chapters. But, as a preview, Holly Kruse wrote in 1995: "A college station with a mainstream format often finds its identity less overtly defined by other college radio stations than it does by commercial stations within its market. When a college radio station is also a commercial station, this relationship to non-college commercial stations is even more noticeable" (168).

Historical Perspective of College Radio

In the academic world we provide what is known as a "review of the literature" when summarizing previous research within an area. Usually the intent is to justify one's own research by proving that it has not been performed before. Also, a review helps to establish the validity of researching a given subject. Here prior research of college radio provides the reader with sources for possible further investigation. By no means are these the only works that have been written on the subject. I truly believe that no overview of the literature is the "definitive" investigation of what has been. In fact, let the following be only the starting point for the reader's search. (Please note that other citations are described throughout this book.)

A prime example of previous research is a study conducted by Garland C. Elmore (1986). Based on the results of a survey of 194 departments offering majors in radio and television, Elmore's research includes data on enrollment figures, facilities, and operating and capital budgets. In an earlier study (1983), Elmore investigated the relationship between administrative organization and radio-television-film academic programs. He surveyed 62.7 percent ($N=131$) of all four-year schools offering a bachelor's degree with broadcasting majors.

Two studies offer insight into college and university radio broadcasting in the 1970s, although limited. One study emphasizes stations in the southeast United States (Drake 1975), whereas the other focuses entirely on carrier-current stations (*College Carrier Current*, 1972).

In a 1984 study of 157 colleges and universities, Gotsch investigated funding for audio and video broadcast laboratories, as well as their affiliation to broadcast facilities. A more focused study on public broadcasting was produced in 1981 from data contained in the *Status Report of Public Broadcasting 1980* of the Office of Planning and Analysis of the Corporation for Public Broadcasting. In this study, funding sources for radio and television were broken down by percentages (Public Broadcasting 1981).

A 1967 study by the National Association of Educational Broadcasters entitled *The Hidden Medium: A Status Report on Educational Radio in the United States* provides a more specific look at the realities of station funding. The report notes that "management, staffing and budget limitations are in the final analysis directly related to school administration attitudes toward the medium," and that, "with few exceptions, institutions of higher education do not accord radio the same degree of concern they do other interests, and thus fail to develop it fully as an educational resource" (113).

In a 1973 study published in the *Educational Broadcasting Review,* Robertson and Yokom focus on the relationships between educational radio stations and parent institutions. They visited 181 noncommercial educational stations to gather impressions of the medium. Factors highlighted include the identification of alternative funding sources such as community fund-raising, underwriting, and listener support. Robertson and Yokom note that "the significance of a public radio station's service in its community is directly proportional to the degree of understanding and support it receives from the top administrators of its licensee institution." They found that "some station managers showed signs of never having been able to talk with their presidents or superintendents" (111).

In a book entitled *The College Radio Handbook* (mentioned in Chapter 1), Brant (1981) points out that "with few exceptions college radio stations are budgeted by the college or university to which they are licensed" (82). Brant also notes that the few commercial stations licensed to educational institutions were able to seek potential advertisers (84). Some of these stations, such as Howard University's WHUR-FM, have even achieved superiority ratings in major markets (Evans 1986). In discussing colleges' and universities' perceptions of college radio, Brant states: "A college that has an AM or FM station sees it more as a public relations tool. Every time some accomplishment is mentioned or the name of the college is mentioned—as in a station identification break—the college's image is enhanced either directly or indirectly" (33). Brant mentions that faculty advisors "act as mediator between the station and the college administration" (143).

Smoot (1988), who describes thoroughly the endeavors of a major university in the establishment of a public radio station, provides actual dollar figures detailing allocation provided by the institution. Other writings have examined surveys undertaken to determine listener interest and programming concerns at college and university radio stations (Rogers 1991, 6).

A study by Bernard Caton (1979) focuses on the aspects of noncommercial radio stations in the Commonwealth of Virginia. In his survey of 23 stations, Caton briefly denotes the purposes served by higher-education radio stations: "Although the purposes of radio stations licensed to institutions of higher education vary somewhat from school to school, several common ones can be identified. Most importantly, all or nearly all stations see their primary function as one of providing alternative programming to their listening audiences"(9).

Two recent dissertation studies focused directly on college and university radio. A study by Poole (1989) offers an in-depth analysis of the marketing positioning theory of public radio stations operated by higher-education institutions. The survey involved a panel of station administrators and experts in public radio who provided opinions on public radio stations' positioning strategies and tactics. The significance of the study is indicated by Poole's statement that "all of the operations involved in development of the product or service, how it is packaged, priced, and presented, should be geared to the goal of customer satisfaction" (24).

Leidman (1985), who examines the state of college and university noncommercial FM radio in a national study, finds that these stations are "almost totally dependent upon the generosity of nonconnected bodies for their funding [and that] they do not generate their own funding either from the selling of advertising nor from charges of tuition and fees" (128). The participants of the study answered questions concerning the specific dollar amounts of their operating budgets. Leidman concludes that percentages within the budgets that come from various funding sources could create an organizationally complicated arrangement (142). Leidman also points out that although NPR stations may be heavily funded by schools as a "sense of mission related to community service," smaller stations also supply a service (238).

Perceptions of college radio are addressed in Reese's 1996 *Feedback* article entitled "College Radio from the View of the Student Staff and the Audience: A Comparison of Perceptions." The premise of the study is geared toward the view of how the college radio station staff and student audience perceive college radio in general. In particular, it is interesting to note respondents' views regarding college radio's primary purpose being that of a training ground.

A recent study of note is that of C. E. Hamilton entitled "The Interaction between Selected Public Radio Stations and Their Communities: A Study of Station Missions, Audiences, Programming and Funding" (1994). Although the study does address the notion of funding, the survey is limited to eight public radio stations to establish the idea of two independent continua (community of service and pursuit of audience) by which public radio stations may be differentiated. Another 1994 study examines the administrative patterns of on-campus radio stations regarding the leadership behaviors of managers. This study (Dennison 1994) is limited to NPR affiliates only.

A dissertation written in 1993 by Charles G. Bailey entitled *Perceptions of Professional Radio Station Managers of the Training and Experience of Potential Employees Who Have Worked in College Radio under One of Three Different Administrative Patterns* provides an outstanding review of the literature (119-24), focusing on college and university radio's employee/student training. As a research tool, a portion of the review is presented here:

Colleges and universities were pioneers in radio experimentation in their science departments, and several of those institutions applied for experimental radio licenses under the Radio Act of 1912. Some of the licensed college radio stations transmitted programmed weather broadcasts, road and market reports, and news before World War I. During World War I, a few of these colleges and universities offered technical radio classes and used their campus radio facilities to train thousands of military and naval personnel to use and operate radio.

After World War I, college and universities that had "broadcasting" licenses issued to a special class of station began broadcasting extension courses. Although 202 AM licenses were issued to institutions of higher education between 1921 and 1936, only 38 were still on the air by 1938 (Frost 1937). This loss of 164 AM college stations may have been precipitated by competition with commercial broadcasters for scarce frequencies and by educational stations' high operating costs and poor quality of programming. The number of licensed college radio stations on-air gradually increased after the FCC's reservation of twenty channels in the FM band for noncommercial educational use.

With the introduction of radio courses into college programs, many college educators began stressing the importance of instruction in radio training, offering suggestions for educating future radio employees, and seeking input from professional broadcasters for the radio curricula. Several of these educators— Brand (1942), Hunter (1944), Tinnea (1947), Williams (1949), and Smith (1964)—provided suggestions and information for designing radio training curricula which would include technical training courses with laboratories and skill courses for all or practically all entry-level positions. The surveys of Whan (1957), Summers (1958), Niven (1961, 1968-69, 1975, & 1986), and Robinson and Kamalipour (1991) identified college and university broadcast educational courses, radio-television programs, degrees, and radio facilities on many campuses for at least forty years.

...A few educators and professional broadcasters like the respondents in Dugas' study (1984) desire employees who are "articulate, knowledgeable, and experienced from a part-time job or internships" (p. 23). Other studies by Higbee (1970), Abel and Jacobs (1975), Rosenbaum (1985), and the Roper Organization (1987) also mentioned extracurricular work or internship in a college or commercial radio station as desirable training for entry-level positions.

Many noncommercial college owned and operated FM radio stations are used as broadcast outlets for educational, informational, and musical programming and as training laboratories. Supervision of many of these college radio stations was primarily by faculty, professional or staff, or student managers in the studies of Rashidpour (1965), Leidman (1985), and Schiller and Schiller (1986). The leadership style of the station managers, according to Oyewole (1972) and Matthews (1978a,b), appeared to be no particular style for often "it depends." Suggestions for determining the role of the campus radio stations and for providing students in college radio with training and professional attitudes for commercial radio were found in the articles by Holgate (1982), Hilliard (1989-90), Thompsen (1991, 1992), McKenzie (1992), and Halper (1992).

Other works are recommended for further insight into the history of college radio. Holly Kruse's 1995 doctoral dissertation *Marginal Formations and the Production of Culture: The Case of College Music* (cited earlier) contains a section entitled "College Radio," which details its impact. (Kruse's work is further examined in Chapter 4.) She writes:

> College radio stations are generally thought of as student-run stations that serve reasonably large campus communities and whose playlists tend to emphasize alternative rock and pop. However, we will see that this stereotype does not apply to many college stations. In order to better understand how college music is disseminated through college radio, it is important to understand how the approximately 1200 college radio stations that exist today fit into the overall radio environment. (158)

Also, as mentioned earlier, Charles Bailey's 1993 dissertation includes a detailed historical background of noncommercial educational radio (42-62). His description focuses on radio and its impact and use within colleges and universities.

Two pieces are of particular note for historical reading: *Education's Own Stations: The History of Broadcast Licenses Issued to Educational Institutions* (1937), by S. E. Frost, Jr., and *The Gas Pipe Networks: A History of College Radio 1936-1946* (1980) by Louis M. Bloch, Jr. These works in particular provide an overview of the historical impact of early college radio.

Finally, ongoing contemporary publications examine the historical values of broadcasting. To get a feel for such scholarly efforts, you may wish to review various works that perform "content analysis" of selected journals. An example of such is Ozmun's *Scholarly but Relevant: A Comparison of Topic Frequency between "Journalism Quarterly," "Journal of Broadcasting and Electronic Media," and "RTNDA Communicator"* (1997).

REFERENCES

Abel, J. D., and F. N. Jacobs. (1975). Radio station manager attitudes toward broadcasting graduates. *Journal of Broadcasting* 19(4): 439-52.

Annual community radio conference. (1996). *Radio World* 20(3).

Avery, R. K. (1998). College and university stations. In *Historical dictionary of American radio,* ed. D. G. Godfrey and F. A. Leigh. Westport, Conn.: Greenwood Press.

Avery, R. K., and R. Pepper. (1979). Balancing the equation: Public radio comes of age. *Public Telecommunications Review* 7(1): 19-30.

Bader, Michael H. (1998). Anarchy on the airwaves: The growing threat of pirate radio. *Radio Ink* 13(7): 52-61.

Bailey, C. G. (1993). Perceptions of professional radio station managers of the training and experience of potential employees who have worked in college radio under one of three different administrative patterns. Ph.D. diss., West Virginia University, 1993. Abstract in *Dissertation Abstracts International* 54/08:2809.

Barnett, R. H. (1987). Auxiliary services: An emerging force on campus. *College Services Administration* 10(3): 30-33.

Bloch, L. M., Jr. (1980). *The gas pipe networks: A history of college radio 1936-1946.* Cleveland, Ohio: Bloch.

Brand, R. C. (1942). The status of college and university instruction in radio training. *The Quarterly Journal of Speech* 28(April): 156-60.

Brant, B. G. (1981). *The college radio handbook.* Blue Ridge Summit, Pa.: TAB.

Broadcasting and cable yearbook 1997. Vol 1. (1997). New Providence, N.J.: R. R. Bowker.

Caton, B. (1979). Public radio in Virginia. Working paper no. 12, Virginia State Telecommunications Study Commission, Richmond. ERIC, ED 183 209.

Cole, H. (1998). Letter of the law: Some thoughts on low-power radio. *Tuned In* 5(4): 24-25.

College carrier current: A survey of 208 campus-limited radio stations. (1972). New York: Broadcasting Institute of North America. ERIC, ED 085 811.

Deal, T. E., and K. D. Peterson. (1990). *The principal's role in shaping school culture.* U.S. Department of Education, Office of Educational Research and Improvement, National Association of Elementary School Principals.

Dennison, C. F. III. (1994). Administrative patterns of on-campus radio stations and the leadership behaviors of the managers. Ph.D. diss., West Virginia University, 1994. Abstract in *Dissertation Abstracts International* 54:3940A.

Drake, H. (1975). A select survey of campus radio stations. Paper presented at the annual meeting of the Southern States Speech Association, Tallahassee, Fla. ERIC, ED 177 633.

Dugas, W. T. (1984, Summer). "Educated" vs. "Trained" production students. *Feedback* 25(3): 22-23.

Elmore, G. C. (1983). The status of broadcast education in institutions of higher learning. *Communication Education* 32: 71-77.

———. (1986). Curricula, budgets, and facilities in education for radio and television. *ACA-Bulletin* 56: 25-35. ERIC, EJ 332 783.

Evans, G. (1986). Howard U.'s radio station rated no. 1 in highly competitive Washington area. *Chronicle of Higher Education* 31(19): 3.

Federal Communications Commission. (1946). *Eleventh annual report: Fiscal year ended June 30, 1945.* Washington, D.C.: United States Government Printing Office.

———. (1948). *Fourteenth annual report: Fiscal year ended June 30, 1948.* Washington, D.C.: United States Government Printing Office.

———. (1950). *Fifteenth annual report: Fiscal year ended June 30, 1949.* Washington, D.C.: United States Government Printing Office.

Federal Communications Commission. (1998). *FCC seeks comments on lotteries, point system to choose between competing applicants for noncommercial broadcast stations.* October 21: Action by the Commission, FCC 98-269.

Frost, S. E., Jr. (1937). *Education's own stations: The history of broadcast licenses issued to educational institutions.* Chicago: University of Chicago Press.

Gotsch, C. M. (1984). University broadcast laboratories: Cost, upkeep, use. Paper presented at the annual meeting of the Speech Communication Association, Chicago, Illinois. ERIC, ED 251 878.

Halper, D. (1992). College radio: A solution to the entry-level job hunt. *College Broadcaster* 6(1): 30.

Hamilton, C. E. (1994). The interaction between selected radio stations and their communities: A study of station missions, audiences, programming and funding. Ph.D. diss., University of California, San Diego, 1994. Abstract in *Dissertation Abstracts International* 54:3025A.

Higbee, A. L. (1970). A survey of the attitudes of selected radio and television broadcast executives toward the educational background and experience desirable for broadcast employees. Ph.D. diss., Michigan State University, East Lansing.

Hill, H. E. (1954). The National Association of Educational Broadcasters: A history. Master's thesis, University of Illinois, Urbana. Copyrighted, mimeographed, and distributed by the National Association of Educational Broadcasters.

Hilliard, R. L. (1989-90). In preparation for professional radio: The role of the college radio station. *Journal of College Radio* 23(1): i, 8-9.

Holgate, J. F. (1982). Determining the role of campus radio. *Journal of College Radio* 19(2): 4, 6-7.

Hughes, K. S. (1980). *A management reporting manual for colleges: A system of reporting and accounting.* Washington, D.C.: National Association of College and University Business Officers.

Hull, R. B. (1956). Consider basic problems. *AERT Journal* (December): 7.

Hunter, A. L. (1944). Education for radio. *Quarterly Journal of Speech* 30 (October): 299-306.

Kruse, H. C. (1995). Marginal formations and the production of culture: The case of college music. Ph.D. diss., University of Illinois, Urbana, 1995. Abstract in *Dissertation Abstracts International,* 56/09:3360.

Leidman, M. B. (1985). At the crossroads: A descriptive study of noncommercial FM radio stations affiliated with colleges and universities of the early 1980s. Ph.D. diss., George Peabody College for Teachers, Vanderbilt University, 1985. Abstract in *Dissertation Abstracts International* 46:1604A.

Matthews, A. (1978a). Contingency management: "It depends." Leadership styles. *Journal of College Radio* 16(1): 2, 4-6.

———. (1978b). Contingency management: Part 2-structure. *Journal of College Radio* 16(2): 16-18, 20a.

McKenzie, R. (1992). A response to Philip A. Thompsen: We can do more to enhance the "electronic sandbox." *Feedback* 33(2): 27-30.

National Association of College Broadcasters. (1995a). *1995 NACB station handbook.* Providence R.I.: National Association of College Broadcasters.

———. (1995b). *1995 college radio survey.* Providence, R.I.: National Association of College Broadcasters.

National Association of Educational Broadcasters. (1967). *The hidden medium: A status report on educational radio in the United States.* New York: Herman L. Land. ERIC, ED 025 151.

Niven, H. (1961). The development of broadcasting education in institutions of higher education. *Journal of Broadcasting* 5(3): 241-50.

———. (1968-69). Eleventh survey of colleges and universities offering courses in broadcasting, 1967-1968. *Journal of Broadcasting* 13(1): 69-93.

————. (1975). Fourteenth survey of colleges and universities offering courses in broadcasting. *Journal of Broadcasting* 19(4): 453-95.

————. (1986). *Sixteenth report: Broadcast programs in American colleges and universities 1986-87* (Broadcast 4623). Washington, D.C.: Broadcast Education Association.

Noetzel, M. S., and J. A. Hyatt. (1990). Auxiliary enterprises: Running a business within an institution. In *Financial management for student affairs administrators,* Media Publication No. 48, ed. J. H. Schuh. Alexandria, Va.: American College Personnel Association.

Ohio House of Representatives. (1969). *Management study and analysis. Ohio public higher education.* Chicago: Warren King. ERIC, ED 031 152.

Oyewole, G. G. (1972). A study of leadership style and effectiveness in educational broadcasting. Ph.D. diss., University of Massachusetts. Abstract in *Dissertation Abstracts International,* 33/12:6945.

Ozmun, D. (1997). Scholarly but relevant: A comparison of topic frequency between "Journalism Quarterly," "Journal of Broadcasting and Electronic Media," and "RTNDA Communicator." Paper presented at the 42nd Annual Convention of the Broadcast Education Association, April 7, 1997, Las Vegas, Nev. ERIC, ED 409 575.

Poole, C. E. (1989). Market positioning [Q8] theory: Applicability to the administration of public radio stations operated by institutions of higher education. Ph.D. diss., University of Florida, 1989. Abstract in *Dissertation Abstracts International* 51: 757A.

Public broadcasting: A status report 1980. (1981). *Educational Record* 62(3): 78-79. ERIC, EJ 252.

Rashidpour, E. (1965). A survey of the present functions and some aspects of organization of educational radio stations in the United States. Ph.D. diss., Indiana University, South Bend.

Reese, D. E. (1996). College radio from the view of the student staff and the audience: A comparison of perceptions. *Feedback* 37(2): 17-19.

Robertson, J., and G. G. Yokom. (1973). Educational radio: The fifty-year-old adolescent. *Educational Broadcasting Review* 7(2): 107-15. ERIC, ED 074 728.

Robinson, W. L., and Y. R. Kamalipour. (1991). Broadcast education in the United States: The most recent national survey. *Feedback* 32(2): 2-5.

Rogers, D. (1991). Finding the right format. *Journal of College Radio* 24(1): 6-7.

Roper Organization. (1987). *Electronic media career preparation study.* Executive Summary. New York: Roper Organization.

Rosenbaum, J. (1985). A college and university curriculum designed to prepare students for careers in nonbroadcast private telecommunications: A Delphi method survey of professional video communicators. Ph.D. diss., Columbia University Teachers College, New York. Abstract in *Dissertation Abstracts International* 46/09:2548.

Runyon, S. C. (1998). FM allocations and classes. In *Historical dictionary of American radio,* ed. D. G. Godfrey and F. A. Leigh. Westport, Conn.: Greenwood Press.

Saettler, P. (1968). *A history of instructional technology.* New York: McGraw-Hill.

Sauls, S. J. (1993). An analysis of selected factors which influence the funding of college and university noncommercial radio stations as perceived by station directors. Ph.D. diss., University of North Texas. Abstract in *Dissertation Abstracts International* 54:4372.

————. (1995). College radio. Paper presented at the 1995 Popular Culture Association/American Culture Association National Conference, Philadelphia, April 14, 1995. ERIC, ED 385 885.

————. (1996). College radio: Points of contention and harmony from the management perspective. *Feedback* 37(2): 20-22.

Schiller, S. T., and S. Schiller. (1986). College O&Os reveal variety. *Radio World* 10(11): 17, 19.

Smith, L. (1964). Education for broadcasting 1929-1963. *Journal of Broadcasting* 8(4): 383-98.

Smoot, J. G. (1988). *The establishment at Pittsburgh State University of radio station KRPS affiliated with National Public Radio and supported by the Corporation for Public Broadcasting.* Kansas: Pittsburgh State University. ERIC, ED 304 114.

Stimson, L. (1998). License auction rules set. *Radio World* 22(18): cover, 14.

Summers, H. B. (1958). Instruction in radio and television in twenty-five selected universities. *Journal of Broadcasting* 2(4): 351-68.

Texas Association of Broadcasters. (1998). Pirate fined, ordered off air. *TABulletin* 42(7): 10.

Thompsen, P. A. (1991). Enhancing the electronic sandbox: A plan for improving the educational value of student operated radio stations. Paper presented to the Broadcast Education Association Annual Convention, Las Vegas, Nev. April 13-15.

————. (1992). Enhancing the electronic sandbox: A plan for improving the educational value of student-operated radio stations. *Feedback* 33(1): cover, 12-15.

Tinnea, I. W. (1947). A radio station manager to teachers of radio. *Quarterly Journal of Speech* 33(3): 334-35.

Welzenbach, L. F., ed. (1982). *College and university business administration.* 4th ed. Washington, D.C.: National Association of College and University Business Officers.

Whan, F. L. (1957). Colleges and universities offering degrees in radio and television: An analysis. *Journal of Broadcasting* 1(3): 278-83.

Williams, H. M. (1949). The status of courses in radio. *Quarterly Journal of Speech* 35(3): 329-33.

Wood, D. N., and D. G. Wylie. (1977). *Educational telecommunications.* Belmont, Calif.: Wadsworth.

3

Programming and Alternative Programming at the College Radio Station

NONCOMMERCIAL FORMATS

Because licenses for noncommercial educational FM stations are the most commonly sought in broadcast college radio, one will find most of these stations somewhere between 88.1 and 91.9 megahertz (within the section of the FM spectrum reserved for noncommercial radio). As a licensee, a "noncommercial educational FM broadcast station will be licensed only to a nonprofit educational organization" and "shall furnish a nonprofit and noncommercial broadcast service" (Rules Service Company 1994-95, pt. 73.503.) Even more recently, nonbroadcast college stations have emerged. These include such outlets as cable-only stations and the concept of Internet broadcasting. Still, these stations usually maintain the noncommercial programming ideals (see Sauls 1998).

Michael Adams and Kimberly Massey's 1995 book, *Introduction to Radio: Production and Programming,* maintains that noncommercial radio can be divided into three categories: college radio, public radio, and community radio. Here each of these categories is discussed for consideration by station advisors.

College Radio. College radio stations dominate the noncommercial channels [*sic*]. These stations typically present *alternative* programming, that is, programs you probably wouldn't hear on other commercial stations. ... Nearly 90 percent of college stations are oriented toward music. ...

In addition to providing unique music and public affairs programming, college radio stations have another primary purpose: to teach students about radio. ... Since college radio allows volunteers to work with any and all departments, students are able to gain first-hand experience in a variety of areas, which provides them with a good "big picture" of the business of radio. (Adams and Massey 1995, 187-88)

(Music programmed at the college radio stations is more specifically addressed in Chapter 4, particularly alternative music.)

It should be mentioned that even college radio stations must adhere to minimum operating schedules. "All noncommercial educational FM stations are required to operate at least 36 hours per week, consisting of at least 5 hours of operation per day on at least 6 days of the week." Concerning holidays, "stations licensed to educational institutions are not required to operate on Saturday or Sunday or to observe the minimum operating requirements during those days designated on the official school calendar as vacation or recess periods" (Rules Service Company 1994-95, pt. 73.561). Stations not meeting these requirements can be subject to share use of the frequency under a time arrangement from another licensee approved by the FCC.

> *Public Radio.* Since public radio stations are noncommercial and cannot compete in the marketplace with their economically more viable commercial counterparts, they are funded partly by the government, by underwriting grants from private corporations, and by listener support. It also helps to be affiliated with NPR because expensive, high quality programming is provided by the network, relieving much of the programming cost responsibilities. The main goal of public stations is to involve the community as much as possible in local, regional, and national issues. ... [As opposed to commercial radio,] revenue received by public stations is put back into programming and operations because public radio is nonprofit. (Adams and Massey 1995, 187)

NPR (National Public Radio) is discussed later in this chapter.

> *Community Radio.* Community radio represents the smallest number of stations in the country, and it usually operates at a lower power level. Stations are economically supported by community groups, local business underwriting, and listener donations. Like college stations, community radio provides alternative programming through a variety of different formats. (Adams and Massey 1995, 188)

Community radio quite often is considered "community access radio"—that which allows the public the opportunity to present programming. Additionally, some would place noncommercial religious stations in the category of community radio.

The point to be made is that no matter what noncommercial format the college or university radio station chooses, it usually can be found within one of these three types. In reality, quite often the format exhibits characteristics found in different combinations of the types. (Of course, you can't discount the idea of putting a commercial station on the air, as is mentioned throughout this book. Some college stations have been very successful in this arena.)

The aspect of "who's listening to college radio" can help to justify the station's format in that the station is theoretically supplying what the listener wants to hear. Thus survey research (as discussed in Chapter 5) is critical at the college radio station.

The fact that the FCC licenses college radio stations to serve their community of license, which usually reaches far beyond the campus boundary, should never be overlooked. This, of course, places an obligation on the station to serve the overall community audience, not just the campus community (see Chapter 8). Thus, giving the audience what it wants, and not necessarily what the students working at the station want to play or hear, clearly becomes a factor—and so programming the station enters the picture.

PROGRAMMING PHILOSOPHY

Programming at college stations "can span many music genres, from rock to folk, jazz to metal, reggae to rap, gospel to tejano, and classical to country. Spoken word poetry, alternative-perspective news, religious and political programming also often find a home on college radio" (Sauls 1995) (see Fig. 3.1). This programming lends to college radio the appeal of the "open radio format." Basically, anything goes. Even the radio drama, the foundation of early radio, can be found today on college radio stations (Appleford 1991) (also see West 1997). As always, programming responsibility rests with the licensee, which in the case of college radio is most likely the school itself. The point to be made here is that total control of the operation of the station, including its programming, can be legally dictated by the school itself (see Chapter 6).

In 1992 a study was conducted that specifically sought to analyze programming elements in public radio. Conducted by Audience Research Analysis and Thomas and Clifford on behalf of the Corporation for Public Broadcasting (the entity that oversees National Public Radio), the study examined 568 stations participating in the survey. The "project's central thrust [was] to seek out underlying patterns in the key dimensions of stations' audience service ... and to identify where these patterns are shared among significant numbers of stations" (Giovannoni, Thomas, and Clifford 1992, 1). The rationale for the study, entitled *Public Radio Programming Strategies: A Report on the Programming Stations Broadcast and the People They Seek to Serve*, was compelling:

> The basic theory of PUBLIC RADIO PROGRAMMING STRATEGIES is that *patterns of affinity exist among stations with shared audience service missions.* By better understanding patterns, the theory continues, public radio can sharpen the effectiveness of the service it provides to traditional constituencies. It can achieve economies of scale through new initiatives that are targeted to meet common objectives. It can reach more broadly across society through programming designed for listeners who are now at the periphery of public radio's audience. (9)

"The *Poconos' New Music Alternative*"
90.3 FM WESS East Stroudsburg University
Fall 1998 Programming Schedule

TIME	MONDAY	TUESDAY	WEDNESDAY	THURSDAY	FRIDAY	SATURDAY	SUNDAY
6 AM / 7 AM	NewsDesk-6:00, Civilization-6:30,World News-7:00, World Business-7:05,Britain Today-7:15, Health Matters-7:30	NewsDesk-6:00, Civilization-6:30, World News-7:00, World Business-7:05,Britain Today-7:15, Health Matters-7:30	NewsDesk-6:00, Sports Intnl.-6:30,World News-7:00,WorldBusiness-7:05, BritainToday-7:15, ScienceExtra-7:30, SportsRoundup-7:45	NewsDesk-6:00, Sports Intnl.-6:30,World News-7:00,WorldBusiness-7:05, BritainToday-7:15, ScienceExtra-7:30, SportsRoundup-7:45	Newsdesk-6:00,Focus on theFaith-6:30,World News-7:00,World Business-7:05,Britain Today-7:15,Founders of Faith-7:30,	NewsDesk-6:00, People &Politics-6:30,World News-7:00,World Business Review-7:05, A Jolly Good Show-7:15,Short Story-7:45.	NewsDesk-6:00, Anything Goes-6:30, World News-7:00,Write On-7:05, InPraiseOfGod-7:15,SportsRoundup-7:45.
8 AM / 9AM	Jestis D's Morning Show (If Jestis can wake up you can!)	Christian Music w/ Josh Clarkson	To Be Announced 100% Commercial Free!			Alternating Currents	Music for Abundant Living with Joan B
10 AM	BBC WorldNews	BBC World News	BBC World News	BBC World News	BBC World News	Jazz for the Common Man with Steve Krawitz *truly* a far-out cat	To Be Announced You could be the next Wess FM DJ!
11 AM	Dylan Hour	DOWN TIME (Your listening flight has been delayed)	When we say we **don't play** "Sleeping Music" **We Prove it!**	Kim's Environmental Talk Show Go Home Plastic foam!	**Random Direction** w/ Andy		
12 PM	Roots and Wings w/				Ever notice how some stations stop the music twice an hour to tell you they stop the music twice an hour!	**Super Sports**	Numa Snyder's
1 PM	John McLoughlin	Christian Rock w/ Ron "Join the listner fellowship today!!"		COUNTRY CROSSROADS Sports Talk!		**Saturday** with Chris!! (Esu may not win on points, but we do occassionaly cover the spread!	Sunday Afternoon Concert Hall
2 PM	folk,celtic,and bluegrass...plus a cool Scottish accent						...a real classical guy
3 PM		In the Light with Mike	Winding Down w/ Mel (from Old Blue Eyes to the Chairman of the board)	The Latino Fiesta w/ Taino	Amy's Alternative Show Progm note: the show's initials are AAS not ASS)	Levi Bradigan Show (No Relation to Laura)	Trip Down Memory Lane w/ Numa "Pennsylvania 6-5000!"
4 PM	Country Crossroads						
5 PM	Latin Music w/ Javier	Mr. Rob's Talk Show	Mike T.'s Hardcore Show! (Hardcore for Hunks)	Ruxpin Radio (Isn't their Brother Famous???)	New York Nightlife (No Beatings, Just cool beats!)		
6 PM	(cuz people like to say Salsa!)	The Stones Hour				Steff and Jenn's
7 PM	Sarah's Buzz Music She sounds so Damn Familar doesnt she?	Fritz's Garage (Just the gold old boys always meaning harm)	Doc Trauma's Operating Theater Where the tools are steralized for your protection!	Smokin Music Network Thats not a beauty rock in his pocket; he's just happy your listening!!	"Left of the Dial" Brings out the best In College Rock!	Ska Show Generic as it Gets w/ Dan (He's bold, bald, and beautiful!!)	Johnathan Park Show! (He Loves Cotton Candy!!!)
8 PM							
9 PM	Hip Hopw/ DJ Gunn! (He's WESS's own Crafty Ninja!)	Hip Hop From The Ground UP! (2 turntables, 2 White Guys and a microphone)	Hip Hop with Lord PK! Program Note: What does PK actually stand for?	The ShadowBox w/ Lori Jane (Stimulating your ears; possibly more)	Mr. Mo's Hip Hop Show! (New Music First!)	If you would like a show Give our Training Director a call at (717) 422-3512,opt 5	TomCrowley: Six Stings more or Lesh! Dead Head Hour from 11-12!
10 PM							
11 PM	Waxin' w/ Jackson! What exactly is he waxing??? Check Please!!!!	Mike Tate's Hardcore Show Hardcore, nothing more then hardcore!	Joe Dembeck	K-D'S Underground Show	WARNING: Exclusive Music not heard on that "OTHER"station in Stroudsburg!		
12 AM							
1 AM	Sick of annoying DJ's?? Have WESS FM DJ your next party! Call (717)-422-3512 Opt 4 for Details. We offer low rates!						Never Any Commericals!!!
2 AM / 3 AM / 4 AM / 5 AM	===============To Be Announced=============== Most likely this time will be filled with the BBC News Service, but some of our DJ's are talented and can go all night long!! And remember our deejays always do it on request because our listeners come first! Call them on the request lines at 422-3133 or 422-3134!! **Over 100 hrs of Live commercial-free programming a week!** NOTE:Our Dj's also promise not to call their wives and kids during their shows!						

1998 WESS FM STAFF The "Homanders In Chief":
Station Manager: Jeff Marsillo (717)-422-3512 Option 1
Program Director: Jason J. Fiore (717)-422-3512 Option 2
Promotions Director: Lori Hartman (717)-422-3512 Option 4
Music Director: Marc Kurtz (717)-422-3099
Training Director: Mike Bidwell (717)-422-3512 Option 5
Thanks For Listening and making 1998 our best year yet!
(Dedicated in Lettice and Lovage memory to Joe Petrucci Now Courier Sports Editor)   *Jason J. Fiore-- PD*

90.3 FM WESS Meets Every other Tuesday at 11am in Stroud Room 203!!

FIGURE 3.1. 90.3 FM WESS Programming Schedule
Source: WESS-FM, East Stroudsburg University. Used with permission.

Many college radio stations put forth programming philosophies to help explain and justify their presence on the airwaves. Quite often these ideologies are contained in station mission statements. An example of a station's programming philosophy as provided via its Web page is presented in Figure 3.2. (WSRN-FM 1991).

SERVING UNDERSERVED NICHES OF THE POPULATION

College students are a frequently underserved audience. Because they often don't maintain permanent addresses, they are skipped by ratings services and their tastes are not accurately measured. (NACB ListServ, October 14, 1998)

Overall, as with commercial stations, the underlying premise of the college radio station is to serve the community, whether it be the campus community or the community at large; but in unique ways, it is often geared specifically to underserved niches of the population. This ideal is consistent with the fact that colleges and universities, like commercial broadcasters, are licensed to "operate broadcast facilities in the public interest, convenience, and necessity"

WSRN's PROGRAMMING PHILOSOPHY

Adopted 29 March, 1991

WSRN: Alternative College Radio. This often-used phrase can lead to misunderstandings about the nature and purpose of Swarthmore's radio station. To help clarify our station's mission, we present some of the most basic points of our philosophy.

WSRN has two objectives which it must fulfill:

- The first is to provide the students of Swarthmore with a chance to work with a radio station. Students interested in communications have the opportunity to become involved with any of the aspects of radio operation and management. In addition, by exploiting the station's record library, students are able to gain in-depth knowledge of rock, jazz, classical, rap, and novelty music.

- The second is to serve the listening community. To maintain our license, it is necessary to provide our outside listeners with programming that expands the range of listening options available to radio audiences.

In order to maintain the largest possible audience, it is necessary for WSRN to provide shows that are not found on other stations. As a result, it is necessary to choose certain types of programs as more desirable than others. The Board of Directors of WSRN does not find its position to be one of judging the quality of different types of music. Rather, it is the board's responsibility to decide which genres of music are elsewhere unavailable. Therefore, students who are knowledgable in non-mainstream music, or willing to learn more about it, are more likely to receive a programming slot than those intent on playing music that can be found on other stations.

FIGURE 3.2 WSRN's Programming Philosophy
Source: WSRN-FM, Swarthmore College. Used with permission.

(Ozier 1978, 34). Studies indicate that the public regards this service to the community as important (see Sauls 1993, Sauls 1995). Research findings reflect "greater diversity in programming and target audiences—not within a given station's program schedule, but among several stations within the same community. For example, even as a given station is working to focus its schedule on a particular 'niche,' other stations in the same community are committed to quite different strategies—different kinds of programming for different kinds of listeners" (Giovannoni, Thomas, and Clifford 1992, 2).

One area being served by noncommercial radio, including college radio, is that of religious broadcasters. Adding to this is the ability of stations now to provide an even greater outreach by delivering their signals to outlying areas via translators and repeaters (as detailed in Chapter 7). Religious broadcasters, a very viable portion of the broadcasting industry (see Fisher 1998), impact college radio within pursuit of the limited number of noncommercial frequencies by other parties, which is further discussed in Chapter 8. As with any programming, religious broadcasts by the college radio station should be previewed to ensure that they meet the station's standards and do not violate Federal Communications Commission rules and regulations. One must remember that the FCC holds the station licensee responsible for all material broadcast by the station. (Information regarding religious broadcasting may be obtained through the Federal Communications Commission Web site at *http://www.fcc.gov/mmb/asd/main/other.html*.) Additionally, equal-time factors can enter into the broadcasting of religious programming. This subject is discussed later in this chapter.

COLLEGE RADIO PROGRAMMING: WHERE DO WE GO FROM HERE?

Thus, as Gundersen wrote in 1989, "much of college radio's charm lies in its unpredictable nature and constant mutations. One fourth of programmers graduate every year. ... No [musical] genre is deemed inappropriate" (5D). Or as Ken Freedman, program director of WFMU, the Upsala College radio station in East Orange, New Jersey, said in 1987: "At best, college radio allows each station to develop its own personality. ... As for us, we're dedicated to diversity—we're specializing in not specializing" (Pareles 1987, C18; see Sauls 1995).

Findings indicate that "individual stations will be more focused in their programming efforts, more discriminating in their program choices. They plan to devote more time to fewer formats. To best serve these stations, producers and funders will need to apply a similar focus and precision" (Giovannoni, Thomas, and Clifford 1992, 2).

90.5 KSHU

PROGRAM SCHEDULE

TIME	MONDAY	TUESDAY	WED.	THUR.	FRIDAY		SAT.	SUNDAY
6AM-12PM	THE CLASSICS	THE CLASSICS	THE CLASSICS	THE CLASSICS	THE CLASSICS	8AM-12PM	SABADO LATINO	FOLK & ACOUSTIC
12PM-3PM	THE OASIS	THE OASIS	THE OASIS	THE OASIS	THE OASIS	12PM-6PM	CLASSIC ROCK	CLASSIC ROCK
3PM-12AM	MODERN ROCK	MODERN ROCK	MODERN ROCK	MODERN ROCK	MODERN ROCK	6PM-12AM	URBAN	URBAN

THE CLASSICS -- *6am to Noon Weekdays* -- Tune in to listen to only the best in Classical music. From Adolphe Adam to Zoltan Kodaly, it's music for the ages.

THE OASIS -- *Noon to 3pm Weekdays* -- Only the best in Contemporary Jazz with music from artists like Luther Vandross, Buckshot Lafonque, Dave Koz and more.

MODERN ROCK -- *3pm to Midnight Weekdays* -- Rock into the night with the best in Contemporary Hits and Alternative Rock from the Smashing Pumpkins, the Wallflowers and Pearl Jam, including the new up-and-comers.

SABADO LATINO -- *8am to Noon Saturdays* -- Tune in to hear the best in Hispanic music from Tejano to Salsa, Caribbean to Baroque in addition to keeping up with issues and affairs affecting the Hispanic community.

CLASSIC ROCK -- *Noon to 3pm Weekends* -- Dust off the vinyl and jam to the best rock from the 60's, 70's and 80's. From the Allmann Brothers to ZZ Top, it's the classic way to enjoy the weekend.

URBAN -- *6pm to Midnight Weekends* -- Enjoy the smooth sounds of Rhythm and Blues, Hip Hop and Rap from Dru Hill, LL Cool J, Blackstreet and more.

FOLK & ACOUSTIC -- *8am to Noon Sundays* -- The only place in Huntsville where you can find the best in Zydeco, Folk and Celtic music as well as a few others. If it needs a home, this is where you'll find it.

Newscasts are at 7am, 8am, 9am, 10am, 11am, Noon, 1pm, 2pm, 4pm and 5pm weekdays.
Sports updates are at 9am, Noon, 4:30pm and 5:30pm weekdays.
Catch Lifetimes and Family Health weekdays at Noon and 3pm.

Listen to local public affairs programming Sundays at 11am and Dialogue at 11:30am Sundays to stay informed of events and people around Huntsville and around the world.

KSHU also covers Sam Houston State University athletic events! Listen to KSHU for more information or check out the KSHU web page (http://www.shsu.edu/~rtf_kshu).

FIGURE 3.3 90.5 KSHU Weekly Schedule
Source: KSHU-FM, Sam Houston State University. Used with permission.

NATIONAL PUBLIC RADIO/PUBLIC
RADIO INTERNATIONAL

> American public broadcasting was established in 1967 as the electronic alterna-
> tive for discussion and documentary reportage on issues of public importance,
> along with cultural, educational, and minority fare. The Public Broadcasting
> Service (PBS) and National Public Radio (NPR) were to create the space that
> the commercial broadcasters thought too unprofitable to provide. They chose to
> manage the "wasteland." (Landay 1996, 19)

Numerous college radio stations are network affiliates or associates of
National Public Radio (or at least carry some NPR-originated programming),
offering well-known programming such as *Morning Edition* and *All Things
Considered.* Research shows these two nationally syndicated programs being
carried by the highest percentage of public radio stations. As of 1996, 70 percent of the public radio stations included in a Corporation of Public
Broadcasting study carried *All Things Considered,* whereas 69 percent aired
Morning Edition (Ryan 1997, 6). (Even if the college radio station does not
carry NPR programs, programmers and managers should at least be aware of
services provided by outside entities such as NPR and Public Radio
International.) Often college stations, modeling themselves after NPR and
community-supported stations, will provide "block programming" along with
"innovative, genre-crossing, free-form excursions" (Pareles 1987, C18). This is
where you will find Gloria Steinem on *City Arts of San Francisco* hosted by
Maya Angelou, or Shire Hite on *To the Best of Our Knowledge* (see Fig. 3.4).
Note that noncommercial formats often include numerous types of music and
programs in "blocks," as opposed to only one music format, as is usually the
case in commercial radio (see Fig. 3.5).

In Chapter 7, funding and the ideals of the Corporation for Public
Broadcasting, which oversees National Public Radio, are discussed. Basically,
NPR operates as a membership organization and a production service, and acts
as the manager of the public radio satellite system, which distributes programming to satellite-interconnected radio stations. For an overview of its satellite
service, I recommend getting a copy of the *Public Radio Satellite System Users
Guide* (National Public Radio 1998b) for how it works, who can use it, and
funding for station satellite equipment. Additionally, the work contains very
useful appendixes, such as a directory of public radio organizations.
Furthermore, stations should obtain a current copy of the *Public Radio Satellite
System Program Catalog* (National Public Radio 1998a) for program listings.

If interested, college radio stations should contact National Public Radio
concerning membership qualifications, associate memberships, program availability, and satellite access to NPR and other public radio programming. The
Public Radio Satellite System Web page is located at: *www.nprsat.org/PRSS.*
Stations that are not CPB qualified can still utilize some NPR programming

through various sources. NPR's main switchboard phone number is (800)235-1212. (Program source listings are described in "Practical Applications" at the end of this chapter.) Concerning programming and CPB, Giovannoni's 1992 study found that:

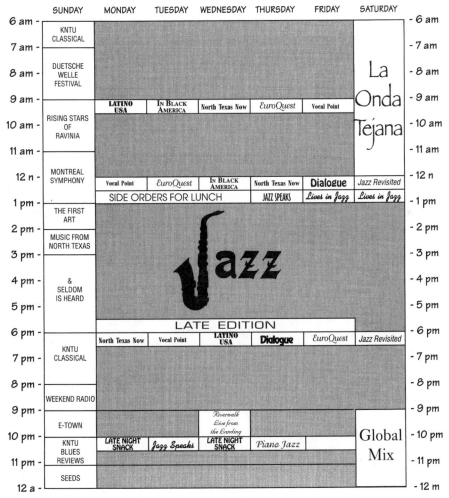

FIGURE 3.4 KNTU 88.1 FM Program Schedule

Source: KNTU-FM, University of North Texas. Used with permission.

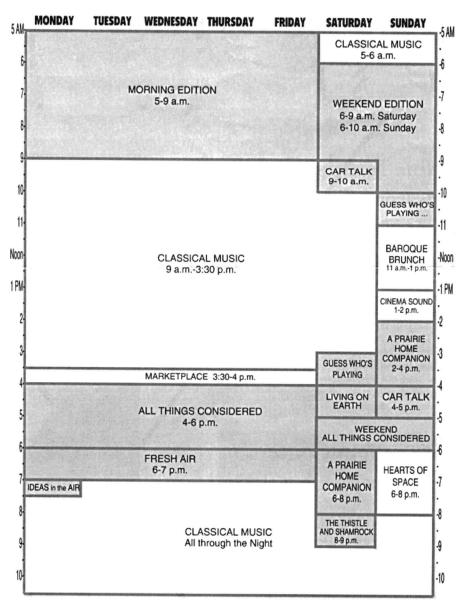

FIGURE 3.5 KNPR Program Schedule
Source: KNPR-FM, Nevada Public Radio Corporation. Used with permission.

There are both differences and similarities between the programming and audience goals of stations that are supported by CPB and those that are not. As "expansion stations" join the system in the next few years they will change it. One dynamic will be new themes in programming and audience targets that these stations introduce. Just as important, if not more so, will be areas in which they align with currently CPB-supported, interconnected stations to create a "critical mass" of stations focused on a particular programming strategy. (Giovannoni, Thomas, and Clifford 1992, 2)

Another programming entity that plays an important role in college radio is Public Radio International. Formerly American Public Radio, the Minneapolis-based company as of 1993 was distributing programming to 475 stations, many of which are also NPR members. "Its offerings include *A Prairie Home Companion with Garrison Keillor* and the widely praised *Marketplace,* a daily business and economic news show. Altogether, APR's programs reach approximately 6.3 million listeners per week, according to Arbitron estimates." (This was compared to roughly 490 NPR member stations and with approximately 1.9 million listeners per week, according to 1993 Arbitron estimates.) (Viles 1993, 52.)

A study of public radio stations carried out by the Corporation of Public Broadcasting, entitled *Highlights of the Public Radio Programming Study, Fiscal Year 1996* (cited earlier in this chapter), found that "Public Radio International (PRI) programming was carried by 87 percent of the stations, surpassing the percentage carrying National Public Radio (NPR) programming (84 percent)." Furthermore, "PRI's carriage reached its height on Saturday evening at 6 p.m. during the live broadcast of Garrison Keillor's *A Prairie Home Companion.* PRI carriage was also strong (30 percent to 40 percent of the stations) during the overnights with their three services: *Classical 24, BBC World Service,* and *Jazz After Hours*" (Ryan 1997, 3). As of 1997 National Public Radio had nearly 540 public radio stations nationwide, whereas Public Radio International had more than 550 affiliate stations throughout the United States, Guam, and Puerto Rico (*Broadcasting and Cable Yearbook 1997,* G61).

WHY RADIO?

Why not? For those of us in radio, and particularly the teaching of radio, we are not only constantly defending our programming but quite often defending radio in general. Then, to add to the overall aspect, we are usually commenting on noncommercial broadcasting. As the roles and responsibilities of those "in charge" of advising college radio stations grow, programming the electronic media entity will become even more important. To help ascertain leadership, guidance, and capabilities of the station advisor (those who foster the efforts of their student staffs), such topics as programming philosophy, general formats, and legal and ethical issues are constantly reviewed (see Sauls 1998). To address public radio in general, Charles Hamilton wrote the following in his 1994 dissertation that provides a true justification for the medium:

Why study public radio?

Public radio is a small and diverse part of the American media. The motivations for the existence of public radio are as varied as anywhere in the American media system. In public radio, we find stations that were founded to explore the science of radio in the 1920s, provide education and culture to the masses, to provide access to media for those who do not usually have access, to train future media professionals, to support political viewpoints, and many more. (6)

PRACTICAL APPLICATIONS

National Programming Interests

The National Association of College Broadcasters' *1995 College Radio Survey* reveals several important programming characteristics of college radio stations and offers several considerations to those who would design a national college radio network.

Seventy-one percent of the respondents reported either extreme interest (21.7 percent) or some interest (49.3 percent) in airing student-oriented national programming. Twenty-nine percent reported little or no interest. Further analysis revealed several differences between these groups.

The majority of respondents with extreme interest (52.1 percent) would air a maximum of two to four hours of national programming each day, whereas the majority of respondents with some interest (71.4 percent) would air a maximum of five or fewer hours a week. Thirty percent of the respondents reported that they have high-quality programs available for national distribution. The most common types of programs offered were specialty, music, and talk programs.

Of great interest to those of us in college radio, the obvious reason for stations expressing an interest in national programming via a satellite network is to fill up those early-morning shows and late, late-night shifts (including overnights) that are hard to staff. Additionally, such programming of a "professional caliber" will serve to augment already locally produced material at the college station. Respondents were asked to indicate their degree of interest in student-produced/oriented music, talk, and special-event programming if such professional-quality programming was made available to them through a satellite network. As Figure 3.6 shows, over 21 percent of the respondents were extremely interested, and over 49 percent reported they had some interest.

Respondents were also asked directly why they had extreme, some, or little or no interest in national programming. Although responses varied greatly, over half of those who reported "extreme" interest, and nearly two thirds of those who reported "some interest," cited that diversity in programming was one rea-

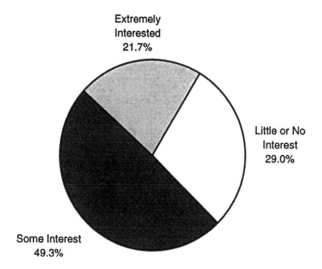

FIGURE 3.6 Interest in National Programming
Source: *1995 College Radio Survey,* National Association of College Broadcasters, p. 33.
Used with permission.

son they were interested in national programming. Lack of need was cited most often by respondents who reported little or no interest in national programming.

National programming desired by stations included special events, news, and radio drama. Again, stations reported that programming available for national distribution included the areas of specialty, music, and talk programs. Table 3.1 presents the operating schedule of stations by number of hours available for national programming. The study shows that those interested in airing national programming would be willing to run commercials or underwriting if it was a required condition for receiving national programming (see Fig. 3.7). Commercial advertising and underwriting are discussed in Chapter 7.

Satellite Equipment Issues

Furthermore, the 1995 NACB study provided data that revealed several important characteristics of college radio stations in regard to the use of programming via satellites.

Specifically for stations, the availability of satellite equipment is of utmost importance. The study showed that, generally, fewer than the majority of college radio stations have their own satellite equipment (37.4 percent). Most of the equipment is C-band equipment (77 percent). Of those respondents who did not have equipment, 61.3 percent have currently nonaccessible equipment nearby.

Finally, respondents who indicated some interest or little or no interest in national programming were asked if their degree of interest would change if

TABLE 3.1   Operating Schedule of Stations by Number of Hours Available for National Programming

Operating Schedule (per day)	Maximum Hours Available for National Programming (*n* = 155)							
	5 or more hours per day		*2–4 hours per day*		*5 hours or less per week*		*Total*	
	Count	%	Count	%	Count	%	Count	%
24–hours	4	8.5	9	19.1	34	72.3	47	100
18–23 hours	5	11.6	13	30.2	25	58.1	43	100
12–18 hours	3	6.3	22	45.8	23	47.9	48	100
Less than 12 hours	0	0.0	7	41.2	10	58.8	17	100
Chi-Square ($p < .11$)								

Source: *1995 College Radio Survey*, National Association of College Broadcasters, p. 67.
Used with permission.

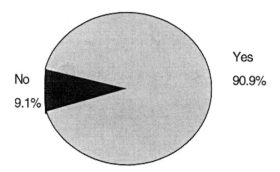

FIGURE 3.7 Willingness to Run Commercials or Underwriting Spots if Required
    Source: *1995 College Radio Survey,* National Association of College Broadcasters, p. 38.
    Used with permission.

they were provided with some of the satellite equipment required to receive national satellite programming. Sixty-seven percent of the respondents reported that their degree of interest would change if the equipment was provided. Eighty-three percent of the respondents with some interest, and over 42 percent of the respondents with little or no interest, reported that their degree of interest would change if the equipment was provided (National Association of College Broadcasters 1995a). (An online source for satellite information is located at *http://www.satcodx.com.*)

Programming Sources

Are there any stations out there who are taking advantage of the recent FCC deregulation of the on-site operator provision to rebroadcast free satellite programming (once consent has been obtained) when that station's local deejays go off the air (usually late night)? For example, let's say your last deejay on a Monday night finishes at 3 a.m. Rather than power down your transmitter, do you switch over to the satellite programming to get you through the night? (NACB Discussion List, March 17, 1996)

The use of satellite programming in the future may be of even more importance as the incorporation of automated station operations expands, particularly with the incorporation of digital equipment (as addressed in Chapter 9). In 1995 the FCC eliminated the need for station operators in order to permit unattended station operation. "The Commission noted that in many areas of broadcast operation, automation is seen as affording more accurate and controlled operation than that performed by humans" (Federal Communications Commission 1995). Here such venues as overnight programming become a true reality, even when dealing with volunteer-student staffs, in that no on-site personnel are required.

Program sources for consideration within the context of a noncommercial venue range from an overnight application up to 24 hours a day and include, for example, *BBC World Service, Classical 24, Jazz After Hours, Beethoven Satellite Network* (via WFMT Networks), *Associated Press 24 Hour News Network, World Radio Network* (foreign news programs), *Radio Blinique* (24-hour Hispanic programming), and *Talk America* (albeit a commercial feed)(NACB Discussion List, March 18, 1996).

The use of satellite programming is becoming even more enhanced as we utilize hard-disk recording systems (such as Audio-Vault at the time of this writing). If the station hasn't yet moved to digital storage, it might consider recording satellite "feeds" on a VCR (videocassette recorder). The signal-to-noise ratio and frequency response are comparable to that of open reel-to-reel recording. But the real advantage is the VCR timer feature to preset recordings (consistent with the operator-free setup already mentioned).

Of course, programming material is available for use via tape and satellite distribution. Syndicated music programs available for noncommercial stations include *The Big Backyard, Spin Radio Network,* and *Left of the Dial* (NACB Discussion List, February 6, 1996). Some student-run stations actually syndicate shows they produce. For example, WEOS in upstate New York syndicated a program called *The Uncle Ziggy Show* and was planning to originate a new show in 1998 entitled the *Nobody Show.* Additionally, WEOS sells its "sports coverage to other stations and produces the Men's Lacrosse Championships for the NCAA" (E-mail to author, October 14, 1998). This is a prime example where stations can originate programming and act as "flagship" stations for other stations, including commercial stations, to receive programming from the college or university.

As of 1998 Randall Davidson, chief announcer at Wisconsin Public Radio, had compiled an exhaustive list of programs available through numerous sources. It is provided in Appendix A to showcase the vast selection of programs available, particularly those open to noncommercial stations by means including traditional (tape, compact disc, etc.) and satellite distribution (see Fig. 3.8).

Of course, the college radio station can't discount the possibility of subscribing to a commercial network for programming, particularly if it is a commercial station. Some provide full-service, 24-hour formats, in addition to short-form programming (see Fig. 3.9).

⇒ **Radio Netherlands now offers to radio stations in North America English and Spanish Language programming on the Galaxy IV satellite for rebroadcast.**

⇒ **FREE OF CHARGE!!**

For Information contact:
LEE MARTIN
Manager of Client Services
Radio Netherlands, North America
1-800-797-1670
lee.martin@juno.com

English & Spanish programming to North America

FIGURE 3.8 Radio Netherlands
Source: Radio Netherlands. Used with permission.

ABC Radio has perfected the art of niche marketing, or targeting the specific needs of an increasingly fragmented radio audience. Local stations can choose from a diverse menu of more than a dozen news networks, including news directed at young adults, contemporary radio, entertainment, urban and rock, and music news from Nashville and in urban music, to name just a few. (Gross 1995, 55)

As previously mentioned, news services should also be considered by the station. Such full-scale services as Associated Press, United Press International, Mutual, USA News Network, Pacifica, BBC World Service, along with other networks, provide "wire" copy (which can be fed directly into a computer-based system), audio networks, and fax services. Any of these services can be utilized to augment news and public affairs programming at the college radio station.

Concerning, particularly, programming supplied by satellite communication, there are critics of the use of the "new technology" in radio broadcasting. Such issues center on the idea that the basic foundation of radio—that of individual image via personality—is being replaced by digitized music and virtual voices and commercials. Here, then, lies the challenge for all in broadcasting to maintain broadcasting through its programming. Digital audio production and digital audio broadcasting (DAB) are both discussed in Chapter 9. Included is an examination of the impact of these new technologies on the programming of the college radio station.

Programming Issues in the Community

To instill the notion that community issues are being presented within the context of public affairs programming, noncommercial stations are required to contact individuals in the local community, usually community leaders, to determine what topics are of current importance. This is where the ideas of taxation, employment, civil rights, infrastructure design, road repair, city services, education, and so on are derived. Within the station public file (discussed further in Chapter 6), a record detailing the contact and programs must be included. The following information should appear in the ascertainment record:

> For nonexempt noncommercial educational broadcast stations, every three months a list of programs that have provided the station's most significant treatment of community issues during the preceding three month period. This list for each calendar quarter is to be filed by the tenth day of the succeeding calendar quarter (e.g., January 10 for the quarter October-December, April 10 for the quarter January-March, etc.). This list shall include a brief narrative describing what issues were given significant treatment and the programming that provided this treatment. The description of the programs should include, but is not limited to, the time, date, duration and title of each program in which the issue was treated. [Effective May 31, 1988] (Rules Service Company 1994-95, pt. 73, record 3662/4224)

The International Advantage...delivering:

Product/Service	Length	Frequency	Form of Delivery
Entertainment			
Award Shows Coverage	various	7-10 shows; yearly	ISDN or DAT
What Else Is News?	various	15-20 cuts; daily	ISDN or DAT
Bill Diehl's Spotlight	60 seconds	5 shows; weekly	ISDN or DAT
StarPower:			
Celebrity Interviews	various	10 interviews; monthly	DAT
Entertainment Minute	60 seconds	5 shows; weekly	DAT
Hollywood Insider	90 seconds	5 shows; weekly	DAT
Music Notes	60 seconds	5 shows; weekly	DAT
Galaxy Artist Newsletter	various	twice; monthly	DAT
Show Prep	various	weekdays	Fax
Holiday Programming	various	December	DAT
Entertainment Weekly Magazine	weekly		Courier
Flexible Music Formats			
24-Hour Formats	24 hours	daily	SAT
Franchise Radio	24 hours	daily	Courier & Fax
Consulting Services	various	daily/weekly	
SoundScan Music Sales Research	various	weekly	Fax
Show Prep Services:			
Morning Show Prep	12-15 pages	weekdays	Fax
ABC's Urban Newsbeat	10-12 pages	weekdays	Fax
Nashville Notes	4-6 pages	weekdays	Fax
RockWire	3-5 pages	weekdays	Fax
Affiliate Conference	3 days	annually; July	
Long-Form Programming			
Rick Dees' Weekly Top 40	4 hours	weekly	CD
Adult Contemporary Top 40	4 hours	weekly	CD
American Country Countdown	4 hours	weekly	CD
Bob Kingsley's Country Specials	various	various	CD
Dick Bartley's American Gold	4 hours	weekly	CD
Tom Joyner's Movin' On	2 hours	weekly	DAT
Doug Banks' Kickin' The Hits	2 hours	weekly	DAT
Z-Rock 50 With Dave Bolt	4 hours	weekly	DAT
Classic Rock Live	1 hour	weekly	DAT
Short-Form Programming			
Daily Features:			
Cybershake	60 seconds	5 shows; weekly	DAT
Science & Technology	90 seconds	5 shows; weekly	DAT
Health Watch	90 seconds	5 shows; weekly	DAT
Travel Around The World	90 seconds	5 shows; weekly	DAT
The Ecology Report	90 seconds	5 shows; weekly	DAT
Business Week Radio Network	60 seconds	19 reports; daily	SAT or ISDN
Court TV Network	60 seconds	daily	ISDN
Black History Month	various	30 cuts; January	DAT
Black Music Month	various	30 cuts; May	DAT

FIGURE 3.9 ABC Radio International

Source: ABC Radio Networks. Used with permission.

▶

Product/Service	Length	Frequency	Form of Delivery
News			
Crisis Coverage	various	various	SAT or ISDN
Special Events Coverage	various	various	SAT or ISDN
ABC Custom Calls	various	various	ISDN
Washington, DC News Reports	30 minutes	weekly	SAT or ISDN
Peter Jennings Journal	2 minutes	weekdays	SAT or ISDN
Newscasts	1-3 minutes	hourly	SAT or ISDN
Special Newscalls	various	10-40 cuts; daily	SAT or ISDN or DAT
Paul Harvey News & Comment	5 & 15 minutes	twice; daily	SAT or ISDN
Paul Harvey's Rest of the Story	5 minutes	weekdays	SAT or ISDN
Audio Datebook	various	80 cuts; monthly	DAT
This Week With David Brinkley	30 minutes	weekly	SAT or ISDN
World News This Week	30 minutes	weekly	SAT or ISDN
Perspective	60 minutes	weekly	SAT or ISDN or DAT
Year-in-Review Programming	60 minutes	December	SAT or ISDN or DAT
TV Magazine Shows Audio	60 minutes	weekly	ISDN or DAT
ABC NewsWire	24 hours	daily	SAT
Sports			
ESPN SportsBeat	5 minutes	weekdays	SAT or ISDN
ESPN SportsBreak	various	various	SAT or ISDN
Extra Point Commentaries	2 minutes	daily	SAT or ISDN
GameDay NFL reports	various	Sundays	SAT or ISDN
Winter and Summer Games:			
Event Reports	60 seconds	19 report; daily	SAT or ISDN
Pre-Event Reports	60 seconds	15 reports; June	DAT
Historical Actualities	various	20-30 cuts; June	DAT
Actualities and Data Feeds	various	6 feeds; daily	SAT or ISDN
Affiliate Support Services			
Technical Services		various	
Sound Services	various	various	SAT or ISDN
Marketing and Promotions		various	
The Walt Disney Company			
Walt Disney World Christmas	4 hours	December	CD
Walt Disney World Radio Studio	various	various	SAT or ISDN

For more information about these products or services contact:
ABC Radio International
13725 Montfort Drive
Dallas, Texas USA 75240

FIGURE 3.9 *(continued)*

At the time of this writing, the rules governing what needs to be included and where the public file is to be located are being reconsidered. As always, current "up-to-date" rules and regulations need to be continually reviewed by station management.

So what does a station program do to address the issues? In sum, the *FCC leaves it up to each broadcaster to determine which issues are of concern to its listening community, and how best to air programs that are responsive to such interests.* This is where it is important for students to stay current with issues in the local community. For students in a college town away from home, this expectation is sometimes quite challenging. Recommendations are for the station public affairs or news director to read the local paper and try to attend local civic gatherings, particularly city council and school board meetings in addition to chamber of commerce events.

Furthermore, those 10-, 30-, and 60-second Public Service Announcements (PSAs) do help in addressing your programming commitment in regard to local issues. "However, PSAs should not be listed as the majority of the responsive programming aired by the station" (National Association of College Broadcasters 1995b, 86; Station relationships with the campus and local community are further discussed in Chapter 8.)

Equal Time

Finally, the question of "equal time" arises. Basically, if one side of a controversial issue is presented, does equal time need to be provided or devoted to the opposing side or sides? Does the station public affairs or news director need to interview all sides when addressing an issue? The Fairness Doctrine —"which required broadcasters to cover issues of public importance and also to provide 'balanced' coverage of such issues"—was repealed in 1987—"based in large part on a Commission study, known as the '1985 Fairness Report,' of the impact of the fairness doctrine on broadcast practices" (Federal Communications Commission 1998). However, as educators (where college radio exists), we are typically professing the ideals of presenting all sides. (See Ken Loomis's 1998 article addressing radio public affairs programming since the Fairness Doctrine.) In reality, however, few in college radio management can foresee fighting the issue of "equal time" in court—we couldn't afford it financially, and our schools may not appreciate it. So in fact "equal time" may actually work against itself.

> *Regarding religious program questions, consider whether you as a station are willing to designate time slots for all possible interested parties that may, in the future, request their own time slots. We weren't and have dropped all such programs. (NACB Discussion List, September 20, 1996)*

There is a belief that if a station is going to be challenged concerning "equal time" allotment, the easy way out is not to present controversial issues at all.

Thus, the ideal of "equal time" has the effect of squelching coverage of issues deemed controversial. Therefore, though the intent of the Fairness Doctrine might have been to present all sides, in use it could actually suppress the coverage of any side by denying broadcast out of fear of reprisal. (Note that political "advertisements" on noncommercial stations are addressed in Chapter 7 within the context of underwriting. Here the context of political coverage concerns discussions of issues by candidates.)

Equal time at the college radio station, like all broadcast outlets, can run the full gamut from public affairs programming to religious broadcasts. Even the ideals of "freedom of expression" fall into this category. A case in point was reported in January 1998 when the Ku Klux Klan in Missouri filed a federal lawsuit against the University of Missouri System and the radio station on its St. Louis campus, seeking to force the station to air the white-supremacist group's slogan. "The lawsuit charged the university and the radio station, KWMU-FM, with violating Klan members' First Amendment rights because the station declined to allow the Klan to underwrite the local broadcast of the National Public Radio program 'All Things Considered' " (Klan Sues U. of Missouri, 1998). As of October 1998 a federal judge upheld that the station did not have to broadcast the KKK announcements (U. of Missouri wins one battle, 1998). The Klan is still pursuing the issue at the time of this writing.

Ethics

In the discussion of programming, ethics must inevitably come into play. The college students who program college radio stations daily must continually consider the degrees of what is right or wrong, funny or unfunny, libel or slander—all these matters come into question in the areas of news, music programming, DJ chatter, public affairs programs, and entertainment/informative interviews. Additionally, personnel and business decisions also involve matters of ethics and ethical practices.

A section of the *1995 NACB Station Handbook* pertaining to ethics was written by Carl Hausman, president of the Center for Media in the Public Interest in New York, a nonprofit media studies agency. Hausman was on the journalism faculty at New York University, is an author of numerous books about the mass media, and has testified before Congress on media ethics issues. He writes:

> While journalists in the United States enjoy Constitutionally-protected freedoms, we realize that from a practical standpoint freedom of the press is not and cannot be absolute. There are various restrictions of an internal and external nature that combine to draw lines on how far a news organization can go in covering a story, or how much privacy must be accorded a subject, or how a wronged person can seek redress against the press. (National Association of College Broadcasters 1995b, 160)

He further states that:

> The word moral is sometimes used synonymously with ethical, although moral-
> ity usually refers not so much to philosophy ... as to prevailing customs. We
> have a tendency to use the word "moral" in matters dealing primarily with
> those customs and not with fundamental questions of right or wrong. For exam-
> ple, we would be far more likely to describe marital infidelity as "immoral" as
> opposed to "unethical." (161)

A recommended reading source, the Hausman piece goes on to address the areas of truth, objectivity, fairness, conflict of interest, sensationalism, and misrepresentation within the realm of journalism ethics. Concerning truth, he states that it is "almost a sacred tenet of journalism, but truth is not always an easy term to define" (163). Furthermore, "an 'objective' reporter is supposed to report 'just the facts' and keep his or her personal opinions out of the piece" (164).

It is incumbent upon station management to instill logic and common sense among the station staff. Good decision making is both inherent and teachable. These attributes will then carry over into sound ethical and moral decisions in both programming and operations.

Program Guide

As a promotions tool for programming, many college radio stations produce and distribute some type of program guide. Word of warning: You are running a radio station, not printing a newspaper! In other words, if you are not careful, you'll end up spending more time producing a program guide than programming your radio station. Program guides can be a lot of work. But they are a saleable piece of self-promotion.

> A program guide should be yet another extension of the station's positioning
> theme, a continuation of the station, in print. Varying from one page, to posters,
> to small magazines, the program guide will only be limited by imagination and
> funds. Selling packages that include print and air-time not only can fund the
> program guide, but also can gently introduce new clients to radio by using a
> medium with which they are familiar. (National Association of College
> Broadcasters 1995b, 203)

Typically, program guides will contain such features as program listings/schedule along with detailed programmed synopsis, current music playlist, sports coverage, special-events coverage (including upcoming station live remote broadcasts), on-air promotions (giveaways), "notes" from the station/general manager, news of past personnel, music artist interviews, and record reviews.

Use the program guide as an in-house morale booster—people love to see their names in print. Pictures are great, as they finally put a face with a name.

As for writing, avoid inside station jokes. The public won't catch what you're talking about and probably doesn't care for station in-house humor. Watch for spelling and grammatical errors. College student writing can be challenging at times. Any college radio station manager should always promote good journalism, both broadcast and print.

Legal Operations

A critical aspect dealing with programming lies in the arena of legal operations at the college radio station. The station/general manager and chief operator (quite often referred to as the chief engineer) are the individuals who are (1) most knowledgeable about station legal matters and (2) responsible for the legal operation of the station. At times the faculty station manager will place the burden of legal operations of the station on the shoulders of a student program director or operations manager. In truth, however, the student is not likely to be aware of the legal concerns of the station. As of 1997 FCC fines (known as forfeitures) ranged from a base amount of $1,000–$10,000 per violation up to a $27,500 maximum per violation, per day, for continuing violations, with a $275,000 statute limitation (see Fig. 3.10). Furthermore, large companies may be allocated even higher fine amounts (McConnell 1997, 19). Additionally, prison terms can also be applied as penalty, depending upon the severity of the station offense.

Keeping up-to-date on legal issues is the manager's and engineer's responsibility. Seeking legal assistance may be necessary at times. When one starts a station, as mentioned in Chapter 2, or when one makes a major change such as a power increase, expert legal guidance will surely come in handy when confronted with ongoing matters. A legal counsel whose specialty is broadcasting (known as an FCC attorney) is needed. Day-to-day lawyers and state attorneys general usually don't speak the "broadcast jargon." And an FCC attorney can save you a lot of time and money in the end!

Areas that are of particular legal concern and consequence for college radio stations in programming are underwriting (including proper log entries and record keeping), legal station identification, contests and promotions, transmitter operation and meter readings, the public file (as detailed in Chapter 6), the EAS (Emergency Alert System), payola and plugola, obscene and indecent material, drug lyrics, and the rebroadcasting of telephone conversations (National Association of College Broadcasters 1995b, 119-21).

EAS should be part of your training. Every DJ should know about it and how to do it. The FCC still makes the licensee responsible for this, meaning your school. ... Every DJ should know how to do an EAS test, just as they did with old EBS. The only difference is the EAS can be set up to do it automatically and log it with ease. (NACB ListServ, September 28, 1997)

Forfeiture Schedule

Violation	Amount
Misrepresentation/Lack of Candor	**Statutory Maximum for Each Service**
Construction and/or operation without an instrument of authorization for the service	$10,000
Failure to comply with prescribed lighting and/or marking	$10,000
Violation of public file rules	$10,000
Violation of political rules: reasonable access, lowest unit charge, equal opportunity, and discrimination	$ 9,000
Unauthorized substantial transfer of control	$ 8,000
Violation of children's television commercialization or programming requirements	$ 8,000
Violations of rules relating to distress & safety frequencies	$ 8,000
EAS equipment not installed or operational	$ 8,000
Alien ownership violation	$ 8,000
Failure to permit inspection	$ 7,000
Transmission of indecent/obscene materials	$ 7,000
Interference	$ 7,000
Importation or marketing of unauthorized equipment	$ 7,000
Failure to maintain directional pattern within prescribed parameters	$ 7,000
Violation of main studio rule	$ 7,000
Violation of broadcast hoax rule	$ 7,000
AM tower fencing	$ 7,000
Exceeding of authorized antenna height	$ 5,000
Fraud by wire, radio or television	$ 5,000
Unauthorized discontinuance of service	$ 5,000
Use of unauthorized equipment	$ 5,000
Exceeding power limits	$ 4,000
Failure to respond to Commission communications	$ 4,000
Violation of sponsorship ID requirements	$ 4,000
Unauthorized emissions	$ 4,000
Using unauthorized frequency	$ 4,000
Failure to engage in required frequency coordination	$ 4,000
Construction or operation at unauthorized location	$ 4,000
Violation of requirements pertaining to broadcasting of lotteries or contests	$ 4,000
Broadcasting telephone conversations without authorization	$ 4,000
Violation of transmitter control and metering requirements	$ 3,000
Failure to file required forms or information	$ 3,000
Failure to make required measurements or conduct required monitoring	$ 2,000
Violation of enhanced underwriting requirements	$ 2,000
Failure to provide station ID	$ 1,000
Unauthorized pro forma transfer of control	$ 1,000
Failure to maintain required records	$ 1,000

FIGURE 3.10 Forfeiture Schedule
Source: Leibowitz & Associates, P.A. Used with permission.

Any station can review the Society of Broadcast Engineers' (SBE) Web site at *http://www.sbe.org* for further EAS requirements. (Transition from the old EBS [Emergency Broadcast System] to the new system of EAS ranged in cost for stations from $1,500 up to $3,000.) Also, stations can view the FCC EAS handbooks on the Web at *http://www.fcc.gov/cib/eas/handbook.htm*. Stations need to be sure to confirm the requirements for their particular status—that is, lower power operation, and so on.

Of course, the best safeguard from legal problems is proper internal station communication and education. Many of these issues can and should be addressed in the station policy manual, which all station personnel should be required to read prior to commencing work as a staff or volunteer member. (Legal issues are further examined in Chapter 6 within the discussion on management.)

Program and Music Scheduling

A final point in regard to programming concerns music/program scheduling and traffic services. With the growth of popularity of the personal computer comes numerous computer programs for the radio station that provide adequate software for music scheduling, traffic (for station program and transmitter/ operating logs), and sales applications that can accommodate tracking.

Because these services are constantly changing and improving, it is not practicable to offer a list of examples. But here are a few points to be made when looking at such programs for the college radio station: (1) It is cheaper and better to purchase industry-tested software than to try to create your own; (2) always ask for an educational discount when purchasing; (3) search for programs that perform multiple tasks; and (4) as always, remember: Whatever you purchase today was surpassed in the design room yesterday (be prepared to update when needed). These programs can also assist the station in operating within legal parameters.

REFERENCES

Adams, M. H., and K. K. Massey. (1995). *Introduction to radio: Production and programming*. Madison, Wis.: Wm. C. Brown Communications.

Appleford, S. (1991). KCRW builds for future with an ear to the past. *Los Angeles Times,* August 28.

Broadcasting and cable yearbook 1997. Vol. 1. (1997). New Providence, N.J.: R. R. Bowker.

Federal Communications Commission. (1995). *Action by the Commission by Report and Order.* FCC 95-214, October 2.

Federal Communications Commission. (1998). *Joint Statement of Commissioners Powell and Furchtgott-Roth.* June 22. (*http://www.fcc.gov/Daily_Releases/Daily_Business/ 1998/db980622/sthfr834.hmtl*).

Fisher, M. (1998). Do 20 million listeners a week spell success for religious radio? *Tuned In* 5(2): 24-26.

Giovannoni, D., T. J. Thomas, and T. R. Clifford. (1992). *Public radio programming strategies: A report on the programming stations broadcast and the people they seek to serve.* Washington, D.C.: Corporation for Public Broadcasting.

Gross, J. (1995). Networking in the 1990's. *Radio World Magazine,* February, pp. 52, 54-56.

Gundersen, E. (1989). College radio explores rock's flip side. *USA TODAY,* February 27.

Hamilton, C. E., Jr. (1994). The interaction between selected radio stations and their communities: A study of station missions, audiences, programming and funding. Ph.D. diss., University of Maryland.

Klan sues U. of Missouri to sponsor show on campus radio station. (1998). Chronicle of Higher Education, January 9, A10.

Landay, J. M. (1996). Don't make public television stations commercial. *Christian Science Monitor,* July 22.

Loomis, K. D. (1998). Radio public affairs programming since the Fairness Doctrine. *Feedback* 39(4): 25-31.

McConnell, C. (1997). FCC issues revised forfeiture guidelines. *Broadcasting and Cable,* August 4.

National Association of College Broadcasters. (1995a). *1995 college radio survey.* Providence, R.I.: National Association of College Broadcasters.

———. (1995b). *1995 NACB station handbook.* Providence, R.I.: National Association of College Broadcasters.

National Public Radio. (1998a). *Public radio satellite system program catalog, spring 1998.* Washington, D.C.: NPR Distribution Division.

———. (1998b). *Public radio satellite system users guide, 1998 edition.* Washington, D.C.: NPR Distribution Division.

Ozier, L. W. (1978). University broadcast licensees: Rx for progress. *Public Telecommunications Review* 6(5): 33-39.

Pareles, J. (1987). College radio, new outlet for the newest music. *New York Times,* December 29.

Rules Service Company. (1994-95). *Radio broadcast services.* Rockville, Md.: Rules Service Company.

Ryan, L. N. (1997). *Highlights of the public radio programming study, fiscal year 1996.* CPB Research Notes, No. 105. Washington, DC: Corporation for Public Broadcasting.

Sauls, S. J. (1993). An analysis of selected factors which influence the funding of college and university noncommercial radio stations as perceived by station directors. Ph.D. diss., University of North Texas, 1993. Abstract in *Dissertation Abstracts International* 54:4372.

———. (1995). College radio. Paper presented at the 1995 Popular Culture Association/American Culture Association National Conference, April 14, 1995, Philadelphia. ERIC, ED 385 885.

———. (1998). Aspects fostering the programming of today's college radio station: the advisor's perspective. Paper presented at the Broadcast Education Association annual convention, April 3, 1998, Las Vegas.

U. of Missouri wins one battle in Klan suit. (1998). *Chronicle of Higher Education,* October 16, A12.

Viles, P. (1993). New name, focus for American Public Radio: As Public Radio International, it will concentrate on the "global viewpoint." *Broadcasting,* December 20, 52.

West, K. (1997). Class radio theatre in contemporary radio. Paper presented at the Annual Joint Meetings of the Popular Culture Association/American Culture Association, March 26-29, San Antonio, Tex.

WSRN-FM. (1991). Home page, Swarthmore College, March 29. *http://wsrn. swarthmore.edu/96/philosophy.html.*

Alternative Music and College Radio

An obvious definition of "college music" would be something like, "the kind of music commonly heard on college radio." For those who share a popular perception of what is played on college radio, this definition brings to mind various guitar-based bands that began and/or remain on independent record labels; bands including mainstream successes like Nirvana, R.E.M., and the Cure as well as lesserknown bands like Paw, Field Mice, and Superchunk. However, college radio stations are anything but uniform in their programming. In the United States today there are college stations with Top 40 or Contemporary Hit Radio (CHR) formats, college stations with album-oriented rock (AOR) formats, and college stations with adult contemporary (AC) formats. Most people who use the term "college music" would not feel comfortable including acts like Mariah Carey, Led Zeppelin, or Boys II Men in the genre. And even college stations that are stereotypically free-format and student-run often have shows devoted to non-mainstream musics like blues, rap, folk, reggae, and bluegrass that tend not to show up on college charts or to be promoted to college radio by record companies. (Kruse 1995, 51-52)

This chapter addresses the role that music plays at the college radio station. More specifically, the "alternative music" format so singular to college radio is discussed in detail. This "alternative" approach is what many believe helps to define each station as a unique entity programming with the consideration of its audience in mind (see Sauls 1998a,b).

ALTERNATIVE PROGRAMMING AND COLLEGE RADIO

Nearly all stations see their primary function as one of providing alternative programming to their listening audiences. ... More specifically, the alternative programming is primarily made up of three types: entertainment, information, and instruction. (Caton 1979, 9)

College radio is as varied as college towns or college students. (Pareles 1987, C18)

Some stations mirror commercial radio, while others opt to develop their own style. Concerning the educational aspect of noncommercial radio (as raised in Chapter 3), some contend that college radio can be educational (for the listener) by programming "diverse" and different music as compared to commercial stations. Additionally, college radio can play a role in "distance education," touted by many academic administrators as a crucial channel for future instruction distribution. In this regard Robert Avery wrote in 1998 that "[b]egun as a distinct counterpart to entertainment-driven commercial radio, which needed to attract the large audience numbers desired by advertisers, educational radio was envisioned as an electronic extension of the high school and college classroom." Avery added that "[b]y the mid-1970s, most of these formal radio courses had disappeared." He concluded that the "principal staples of educational public radio had become public affairs programming and classical and jazz music" (Avery 1998, 135).

Geography also helps determine a college station's format. A college station in rural Indiana might be the only station on the entire dial supporting the Foo Fighters, but if a college station based in Chicago plays the same band, it is likely playing the same music as five other commercial stations. (Marcus 1997, 27)

Some have defined alternative to be something not played in the market. In some cases, this could mean classic rock, country, etc. However, when one thinks of college radio, it usually features new artists, independent bands, and things you won't hear anywhere else. Advantages: Diversity, always new, gives artists and listeners exposure. Disadvantages: Constantly reinventing itself, possible listener tuneout, if too diverse, continual argument of what the listeners want to hear vs. what you think you should be playing. (NACB Discussion List, October 15, 1996)

The programming of alternative music can have a negative impact as well. This concept of college radio as an alternative to commercial radio is fairly widespread, but as Thompsen indicated in 1992:

It can detract from the educational experience of students by encouraging them to focus on the sources of programming, rather than on the audiences for programming.
   ... The philosophy is, by design, diametrically opposed to the prevalent philosophy of nearly every commercial radio (and television) station. (13)

In reality, the entire concept of providing "alternative music" to a college audience can be questioned as to the penetration of the college demographic itself. Kevin Zimmerman wrote in 1989 that "more high schoolers actually listen to alternative music than college students" (67). And so it is postulated that although alternative rock bands are popular on college radio stations, the college students themselves listen more to mainstream radio. "Some program directors argue that college stations with mainstream formats better prepare broadcasting students for careers in commercial radio. Others cite the nature of the audience" (Kruse 1995, 167-68).

As noted in Chapter 2, a case study focusing on alternative music, particularly in conjunction with community and college radio, is presented in a dissertation written in 1995 by Holly Kruse. Entitled *Marginal Formations and the Production of Culture: The Case of College Music,* the work highlights three locations: Athens, Georgia; San Francisco, California; and Champaign-Urbana, Illinois. Additionally, the work provides a good historical perspective concerning college radio. Kruse offers this caveat: "When looking at how college radio fits into both the overall radio market and specific local markets, it is tempting to conceptualize college radio as a monolithic medium. This temptation should be avoided, however, because not all college radio stations follow alternative music formats, and college radio is not the only form of radio that programs alternative music"(Kruse 1995, 167).

As addressed in Chapter 2, the growth of noncommercial educational FM radio, generally the staple of college radio, can be attributed to the Federal Communications Commission's allocation in 1945 of twenty FM channels designated for noncommercial use (between 88 and 92 megahertz). Few anticipated that the noncommercial "band" would evolve into what it is today. "In the later 70's and early 80's, bands like R.E.M., U2 and Talking Heads first established themselves on the underground circuit before eventually reaching an audience of millions" (Schoemer 1992a, 26). This underground programming could be found on college radio.

Back in the '70s it would probably have been rebellious playing a Black Sabbath record or maybe the Sex Pistols. These days it's going to be something like the Butthole Surfers. The line keeps getting moved further and further to the left. Certainly to someone not really involved in it, it could seem like "it's much more out there now than it was." But in a historical context, and in taking things as they relate to each other, it's just as outrageous to play a Black Sabbath record in 1970 as it is to play the Sex Pistols in 1976 as it is to play the Butthole Surfers today. It still achieves the same effect. (G. Gimarc, author of Punk Diary: 1970-1979 *and* Post Punk Diary: 1980-1982, *personal communication, March 8, 1995)*

When did the alternative music "thing" really take off? In 1994 Cheryl Botchick, an associate editor at the College Music Journal New Music Report, said that "[t]en years ago, college radio existed in kind of a bubble. ... Then came Jane's Addiction, Nirvana, Lollapalooza, Pearl Jam, and the lucrative marketing of alternative music" (Knopper 1994, 84). Some record companies have gone so far as to suggest "that college stations are 'wasting their signal' if they aren't playing alternative music" (Stark 1993, 90).

Why has "alternative music" become so popular? It is projected that some 70 percent of all campus radio stations licensed to colleges and universities program some type of "alternative rock" (Wilkinson 1994). Radio consultant and Pollack Media Group CEO Jeff Pollack said that "[p]eople are taking a rawer, tougher, more substantive approach to things in general, and there's a rejection of what's predictable and too slick" (Zimmerman 1992, 66). This feeling has also thus produced talent like Soundgarden, Stone Temple Pilots, Tracy Chapman, Edie Brickell and the New Bohemians, Living Colour, Ziggy Marley, Dire Straits, the Police, the Cars, the Clash, Elvis Costello, 10,000 Maniacs, and Nine Inch Nails (whose Trent Reznor is known as Marilyn Manson's mentor—the "studiously abrasive-noise metal band"[The usual suspects, 1998, 44]). These are bands that became commercially viable (now well-known) "mainstream" groups.

But, as noted earlier, alternative music is not limited only to rock music. In 1989 Gil Creel, music director of Tulane University's WTUL, New Orleans, told the audience at the College Music Journal's New Music Report's Music Marathon that "concentrating on 'the latest kick-ass hardcore or feedback [rock]' wasn't enough to be alternative, but that jazz, house music, hip-hop, and blues must also be represented" (Bessman and Stark 1989, 12). Even "college-appropriate country music" should be exposed (Bessman 1989, 52).

And these formats are legitimate alternative contenders on noncommercial stations. "Among CPB-supported stations, jazz is one of the dominant music formats, airing on 67 percent of the stations and accounting for 17 percent of the total weekly hours of broadcasting, according to a Spring 1994 programming survey" (Gronau 1995b, 22; see also Gronau 1995a.) Furthermore, a study by the Corporation for Public Broadcasting found that "two thirds of the stations' broadcasts were music based, and classical music or jazz described three fourths of the music broadcasts. Jazz based programming was found on more schedules (86%) than classical base (75%); however, the weekly average of classical music (69 hours) was more than twice the weekly average of jazz (29 hours)" (Ryan 1997, 2).

"College radio is a safety valve in the sanity of the music world" (G. Gimarc, personal communication, March 8, 1995). This is where new talent is born and discovered. "Today, college radio is all-important. It's the breeding ground for the new talent ... [and] it's also the lifeblood of the independent record industry" (Ward 1988, 47). By the mid-1980s it was discovered that college radio could break new groups in such genres as country/punk fusion, the '60s sound, and punk rock. "College radio stations—greenhouses for cutting-edge rock 'n' roll—nurture new bands that often become mainstream hits a few months later" (Stearns 1986, D4). Thus, "[a]t a time when many new artists face difficulty breaking through at commercial radio, college radio has grown into a virtual industry within an industry" (Starr 1991, 30). Major music industry trade magazines, such as *Gavin,* now cover college radio playlists, along with mainstream music (see "Charts and Music Reps" in this chapter). "The major record companies view format-free college stations that play alternative music as rock's minor league, the training ground for future U2s and Depeche Modes." Additionally, college students playing the music tend to appreciate it more than mainstream DJs. "Says singer Tanya Donelly, who's often interviewed by campus deejays, 'They're more educated and excited about the music.'"(Mundy 1993, 70). This leads to the social implication of discovering new talent. "College radio is garnering new respect and clout as a launching pad for undiscovered, and under-appreciated, talent" (Gundersen 1989, D5). New talent, alternative music, and college radio are more and more in demand (Mayhem 1994).

COLLEGE RADIO AND THE MUSIC INDUSTRY

And college radio has made an impact, and the music industry has discovered college radio. (See Holterman's 1992 thesis, *The Relationship between Record Companies and College Music Directors: A Descriptive Study of Alternative Radio.*) In October 1992 Schoemer wrote:

> The music industry at large has looked for ways to exploit college radio as market at least since the mid-80's, when bands like R.E.M. and U2 crossed over from a base of college-radio fans to mainstream commercial success. But this year [1992], with the multi-platinum sales of albums by Nirvana (more than 4 million copies), Pearl Jam (over 3 million copies), Red Hot Chili Peppers (3 million) and others, the game has changed considerably. College radio has been a business for several years; now, it's serious business. (1992b, C27)

The music industry is turning to "college radio as a kind of early warning system, identifying bands that may reach mainstream audiences an album or two in the future" (Pareles 1987, 18). "Major labels access college radio most directly through alternative music promotion departments which began springing up at majors during the eighties and which exist almost exclusively to get records played on college radio (and, to a lesser extent, commercial alternative radio stations)" (Kruse 1995, 166).

> Individuals at college radio stations, however, unlike most people at independent record labels, tend to stress the important roles played by locality and local music in the constitution of their stations' identities. Obviously a radio station does not have the geographical scope available to even a small independent label, which can reach national and international audiences through independ-

ent distribution and mail order; college (and community) radio stations are rarely able to reach audiences beyond a ten to twenty mile radius of their transmitters. College radio is an inherently local medium, and college (and community) radio stations, to varying degrees, articulate their identities as constituted by locality. (Kruse 1995, 166)

(Charting college radio music and music industry representatives are discussed within the Practical Applications at the end of this chapter.)

IT'S NOT MAINSTREAM ... IT'S ALTERNATIVE ... IT'S EXPRESSION

If the *Saturday Night Live* appearance was Nirvana's chance to prove to the unconverted that it was worthy of such honors [of being one of the most popular bands], the group failed miserably. But if its goal was to make an uncompromising display of the values of underground music, the achievement was unheralded. Nirvana didn't cater to the mainstream; it played the game on its own terms. (Schoemer 1992a, 26)

For those of us who saw the Nirvana television performance, it was a message of expression. Here the ultimate alternative, college-oriented group was captivating and cultivating the American culture with its performance. For them it was their time to express their true feelings. "Nirvana may not fit into the formulaic pigeonholes the industry usually carves for popular music. But for a whole generation of misfits, the members of Nirvana are nothing short of saviors" (Schoemer 1992a, 26). Coupled with Nirvana's appearance on MTV's *Unplugged*, this performance was the true definition of the young culture and their feelings toward society. "Generation X" has realized that they might not be as successful as their parents. This phenomenon, if it becomes reality, will be a societal first in America. We have always strived for, been preached to and told about, the certainty that we will do better than our parents. But that certainty no longer exists—and the college generation, the younger generation, knows it. To other generations they appear to wander about, misguided. In a way, their response to life is almost an organized form of being hectic. ("GenXers," nevertheless—like their predecessors—are organized to the extent of putting together large concert events. See Associated Press 1998a.) You really don't know what's next, but you can sense the anger. And this form is what is reflected in their music. In turn, it is reflected in the programming of college radio (see Esselman 1996).

Programming on college radio can be directed solely to this generation. "It's like 'Wall Street Week' for Generation X," a 1998 article marvels. "On a radio show hosted by a Northwestern University senior, the price of marijuana is used to illustrate supply and demand, and female business journalists are described as 'finance hotties'" (Note Book 1998, A52). The radio program,

Capitalist Pig, is targeted at a twenty-something demographic and is provided by Public Radio International's *Marketplace.*

Incidentally, it is interesting to note that Generation X is the "cleaner generation." A survey by Teledyne Water Pik reports that Generation X "takes longer showers than boomers. Average time: 16.4 minutes. Note: Women shower 2 minutes longer than men" (Pitts 1995, 6).

WHAT'S THE FUTURE OF ALTERNATIVE RADIO?

As of September 1994, the *M Street Journal* reported that out of 11,565 operating stations, 370 were programming alternative rock as their primary format. Of these stations 276 were noncommercial. Even more influential, alternative rock was listed as the tenth-most-popular format out of twenty-nine, indicating its growth as a fundamental format (October sample, 1994). Thus I believe we can safely project that alternative music will continue to maintain, if not expand, its influence on college radio stations (while continuing to enter formats of commercial stations, as well). These numbers clearly indicate that alternative music is a very viable part of noncommercial radio. Add to this the fact that, with "more than 1,100 college stations to appeal to, many bands just breaking into the music industry can be heard across the United States" (Allen 1997, 13).

The 1992 study by Audience Research Analysis and Thomas and Clifford, conducted on behalf of the Corporation for Public Broadcasting (see Chapter 3), specifically sought to analyze programming elements in public radio. With 568 stations participating in its survey, the "project's central thrust [was] to seek out underlying patterns in the key dimensions of stations' audience service ... and to identify where these patterns are shared among significant numbers of stations" (Giovannoni, Thomas, and Clifford 1992, 1). In the study it was established that local programming is paramount when identifying alternative stations, particularly those which promote localism. Here, the proposition that they are "alternative" to other radio stations in their market is professed. The study, when addressing the future of alternative programming, notes that these stations exhibit what the authors call "intracohort diversity" (22). This philosophy helps to bring forth the idea that distinctive programming needs and ideals are present that could be applied or at least accepted among the bulk of these types of outlets.

A final thought concerning alternative programming in college radio and its future:

You get a lot more confrontational radio in college radio. And it's actually very healthy from two different standpoints. One is it gives an accurate reflection of that part of society which doesn't really get much exposure in the normal media. And the other part is, as broadcasters, these kids get to work it out of their systems for four years before they have to put on the suit and tie and be real people ... real responsible broadcasters. (G. Gimarc, personal communication, March 8, 1995)

Some believe that "college radio is divided between innovation and complacency. Some insiders believe college radio is still paving the way for new artists and new sounds" (Marcus 1997, 26). As discussion continues on the subject of college radio, one thing is clear: Just being different is no justification for a format. Managers and programmers must always consider their station mission and intended audience. In any case, the programming of "alternative" formats will continue to play a vital role in the medium that is college radio.

ALTERNATIVE MUSIC AND FREEDOM OF EXPRESSION

Throughout this chapter the ideals of expression and the idea of censorship must be considered within the discussion of alternative music and its relationship with college radio programming. From the outset, the right of free speech must be addressed.

PROTECT YOUR FREEDOM.*

Freedom of the press in this nation was guaranteed by our founding fathers when they penned the First Amendment. Since that time, the Fourth Estate—so dubbed by Edmund Burke in 1841 as one of the components of the British realm along with clergy, nobility and commoners—has enjoyed a privilege and stature envied by many nations and used by them as a model for their own fledgling democracies.

It is ironic that this paradigm of freedom does not extend to the electronic press. Content regulation of television and radio—in the form of "safe harbor hours" and V-Chips and possibly even an A-Chip—is, by extension, unconstitutional.

Our governmental system of checks and balances failed the broadcasters recently, when the Supreme Court of the land upheld a "safe harbor" set forth by the U.S. Court of Appeals last summer.

With the high court's endorsement, the Federal Communications Commission can enforce the 6 a.m. to 10 p.m. ban on the broadcast of indecent material.

Never mind that now, as one group of broadcasters put it: We are "forcing adults to see and hear only programming that is suitable for children."

Consider the implications for any new technologies in development. The sheer multiplicity of electronic distribution pipelines in existence now and soon to come reinforces the argument that First Amendment protections must be

---

From Radio World, February 7, 1996, p.5. Used with permission.

equally applied to electronic and printed press. The distinction is getting harder to see by the minute.

The question begging to be asked is: Will an uncensored Washington Post or Playboy magazine, for example, need to be censored for distribution on the electronic pipeline? And if so, why?

The pioneers of electronic media, including radio and television broadcasters (who have, after all, only been around a short 75 years) will have to make a stronger effort to protect their rights if First Amendment parity with the printed press is to be achieved.

PROGRAMMING, CENSORSHIP, AND ALTERNATIVE RADIO

"From scholarly jazz programs to unusual classical repertory to crashing, howling post-punk hardcore rock, college radio (alongside a few listener-supported and community radio stations) supplies music heard nowhere else on the airwaves" (Pareles 1987, 18). This type of programming, an actual service, is consistent with the fact that colleges and universities, like commercial broadcasters, are licensed to "operate broadcast facilities in the public interest, convenience, and necessity" (Ozier 1978, 34). Additionally, the ongoing broadcasts provided by college radio help to serve as public relations arms for the schools themselves. Often college radio stations are the only outlets for such broadcasts as campus sports and news. In regard to the colleges' and universities' perceptions of college radio, one advantage is that the institutional image is enhanced every time a well-programmed station identifies itself as affiliated with the school (Sauls 1995). This identification, though, can lead to potential problems, particularly in regard to music programming and censorship. Wolper, in 1990, clearly indicated this issue when he wrote that "[t]he licenses of campus radio stations are held by boards of trustees at universities and colleges. Those groups traditionally avoid arguments with the FCC" (54). Wolper cited the concern of Ken Fate, the student general manager of KUOI-FM at the University of Idaho in Moscow, in regard to the FCC: "They are trying to censor us. ... They are making it criminal to play music. To read poetry on the air. To read literature" (54).

> As the influence of college radio grows . . . so does its caution. [In 1987] the Federal Communications Commission issued a warning to KCSB-FM, a 10-watt college radio station in Santa Barbara, Calif., that it had committed "actionable indecency" by broadcasting the punk-rock song "Makin' Bacon" after 10 p.m. The warning was part of a broadening of the commission's restrictions on broadcast indecency. (Pareles 1987, 18)

Thus, the issue of censoring the programming of alternative radio, and radio in general, is introduced. Because alternative music tends to "push the limits" at times, it will, of course, be subject to more scrutiny. Care must be taken to ensure that "clearing" music for airplay does not cross over into outright censorship.

But managers and programmers must be able to ascertain what could be deemed as objectionable within the areas of obscenity and indecency. To aid in clarifying this dilemma, two discussions are presented in this chapter. The first considers "expositional obscenity," particularly in the role that college radio plays in establishing boundaries of acceptable broadcast expressions. Then, within the Practical Applications, dealing with obscenity and indecency in actual station program operations is discussed. Included are fundamental definitions of obscenity and indecency and how they are interpreted within noncommercial stations.

THE ROLE OF EXPOSITIONAL OBSCENITY IN COLLEGE RADIO

Robert McKenzie, Ph.D.
January 24, 1994
Associate Professor of Communication Studies
University Advisor to WESS Radio
East Stroudsburg University

Presented at the 1993 SCA Convention in New Orleans.
Used with permission.

(This paper was originally titled "The Role of Obscenity in College Radio" and first presented at the Freedom of Expression Division of the 1993 Speech Communication Association convention in Miami, Florida.)

Obscenity is a relative term. It means—in a very loose kind of way—any behavior (including language) that some group of people finds to be extremely offensive. Obviously, the definition is problematic because some people think the expression "fuck you" is offensive, while others think the expression "abortion kills" is offensive. The constitution of the United States was amended to guarantee the free exchange of potentially clashing ideas as "protected speech," but the Supreme Court has ruled that obscenity is a realm of public discourse that falls outside the boundaries of protected speech (U.S. v. Roth, 237 F.2d 796, 1956).

Broadcast media are not free from such restrictions either, as the Federal Communications Commission has been empowered by Congress to revoke a radio or television station's license for operation as a result of an obscene broadcast *(Code of Federal Regulations:* Section 1464, Title 18). Despite this provision, however, material that some would consider to be obscene has been regularly aired by broadcast media, and no radio or TV station to date has been denied renewal of its license for broadcasting obscene material (Pember 1984).

This paper is an examination of obscenity in college radio. My thesis is that college radio uniquely interacts two ends of a language spectrum to produce a catalytic form of discourse I call "expositional obscenity." I argue that expositional obscenity in college radio invites the audience to render a judgement about questionably obscene material. After the judgement is rendered, it is adapted [ ] by conventional radio and television stations. Three factors contribute to my thesis: (1) certain laws and guidelines that regulate the use of obscene speech in broadcast media, (2) obscenity in college radio, and (3) the progressive role of obscene college radio in expanding the boundaries of conventional broadcast discourse.

Federal Laws and Guidelines

Three legislative acts are responsible for the way the federal government regulates speech in broadcast media today. All three acts have been set up under a philosophical outlook in the United States that generally views radio and TV stations as guardians of the public's welfare; the philosophical outlook is the so-called "public trusteeship" model of broadcasting (Sterling and Head 1990). A critical assumption of the model is that broadcast media should serve the public's interest because irresponsible media programming can cause mental or physical harm to the audience.

The Radio Act of 1927 put the public trusteeship model into law by specifying that radio stations should serve the public's "interest, convenience and necessity." To interpret this broad guideline, the Act brought regulation of broadcasting under the control of the federal government. As part of the Act, the Commerce department was given the power to license all radio stations in the country; and stations that did not demonstrate that they would serve the public's interest, convenience and necessity would not be licensed.

Obscene language during this time period was virtually nonexistent in radio programming for two related reasons: (1) If a station broadcast obscenities it might not get its license renewed; and (2) the predominant source of funding for radio was advertising by single-sponsor programs unwilling to have their products associated with "dirty" programming. Neither of these reasons had very much to do with actual needs of the audience.

The Radio Act was modified by a second piece of legislation, the Communications Act of 1934, which formed the Federal Communications Commission (FCC). This new streamlined federal agency was now authorized to oversee the licensing and regulation of all telecommunications media, which included radio, the telephone, the telegraph, and, for a new technology at that time, television. The far-reaching Communications Act put into law the specific procedures by which radio and TV stations should serve the public's interest, convenience and necessity. The procedures covered technical configurations for radio and TV stations, public input in a station's licensing process, and even categories of content a station must broadcast.

As a result of the Communications Act, public input became such an integral part of the FCC's licensing process that stations could actually be prevented from abandoning an unprofitable format if there was no similar format around and the community argued strongly enough for retaining the format. Other regulations were just as prescriptive. For example, advertising was limited to eight minutes per half hour, news and public affairs were stipulated to be at least 10 percent of a station's overall programming, and station identifications were required at the top of the hour.

In the era of the Communications Act, obscenity on the radio gradually began to emerge as a controversial issue. The emergence of obscenity on radio came about as the industry moved from single-sponsor programming to multiple commercial advertisers, a situation that created an atmosphere of competition for product recognition within single radio and TV programs. In other words, back-to-back commercials loosened the monopolistic control that sponsors previously had over creative programming, and words like "damn" and "hell" began to be heard on the air as natural expressions of deejays seeking to entertain their audiences.

Broadcasting continued to test and expand the boundaries of obscenity until the Supreme Court upheld a case in 1978 that prohibited the airplay of specific words. In Pacifica vs. FCC, the Court ruled that George Carlin's "seven bad words you can't say on TV," were not permissible on the air; other potentially obscene words of more ambiguous meaning like "ass" continued to be heard on the air. In other words, the court ruled that not all speech in broadcasting was "protected speech." Furthermore, a violation of this ruling would constitute grounds for the FCC to fine a station or deny renewal of its license.[1] Thus for the first time, broadcast language was cast outside of the rights provided to individuals by the First Amendment.

However, in the 1980s the FCC under the direction of then-Chairman Norman Fowler took a decidedly different view of obscenity by deregulating the procedure for determining whether a word is obscene. The deregulative approach to broadcasting was to let the audience—defined as a marketplace—decide what was to be considered obscene programming (Fowler and Brenner 1982). If the marketplace voiced enough of a concern over potentially offensive programming, and documented the alleged obscenity with either transcripts or tape, the FCC would step in with a fine. As a guideline for determining the validity of a public complaint, the FCC defined obscenity as "language that describes, in terms patently offensive as measured by contemporary community standards for the broadcast medium, sexual or excretory activities or organs" (Donnerstein, Wilson, and Linz 1992, 111).

Thus, by shifting the burden of determining what is obscene from the FCC to the public, deregulation became the third major legislative activity to affect the issue of obscenity in broadcast media. The effect of deregulation should not be underestimated since it put into motion a fundamental shift in philosophy about the role of broadcasting; namely, it changed the role of broadcasters as trustees of the public good to broadcasters as commercial distributors of products responding to consumer needs of the audience. And while the FCC still retains the right to fine broadcasters on a case-by-case basis for obscene programming, deregulated broadcasters today are actively exploring programming in a way that challenges the FCC's traditional view of obscenity.

Unfortunately, broadcasters cannot really be sure what the boundaries of obscenity are until a fine has been received from the FCC, so deregulation has created a climate of legal uncertainty (*Broadcasting,* March 9, 1992). While Carlin's "seven bad words" are still generally viewed by the FCC and broadcasters alike as off limits, a flurry of new expressions on the air is presenting murkier territory. Words that have emerged today as tests of the obscenity boundaries include: penis, butthead, ass, boobs, and nipples.

During the transitory period of deregulation, TV shows like *Married with Children* and *NYPD Blue* have challenged the boundaries of obscenity with visual images, but certainly radio has been the more daring broadcaster. Most people know of the fines levied by the FCC against the Infinity Broadcasting company for obscene scenarios described on its Howard Stern show, but many regional and local radio stations have also been fined in recent times for obscene broadcasts (*Broadcasting,* February 22, 1993; *Broadcasting,* March 2, 1992). One station, WLUP, was fined for the obviously tasteless deejay joke asking: "What do you do after you eat a bald pussy? Refasten the diaper."

Leaving aside the sickness of the joke as well as a discussion about whether it promotes or creates social deviancies, the joke nonetheless shows the risk broadcasters are willing to take today to entertain or shock the audience. Although WLUP was fined, the fact that lawyers for the Howard Stern show have challenged their FCC fines in court, and that no fines have been paid as of yet, illustrates how cloudy the issue of obscenity in broadcast media has become. More important, the Stern-FCC standoff reveals a peculiar quandary in conceptualizing what on-air material can be proven as obscene; Carlin's words are obscene because they are discreet symbols, but Stern's scenarios may not be obscene because they are audience imagined.

Obscenity in College Radio

Nowhere in the broadcast media is the climate of obscenity more malleable than in college radio, which provides a unique combination of elements where its deejays invoke two ends of the obscene language spectrum to explore new forms of obscenity. The combination of elements unique to college radio are: (1) youthful deejays experimenting with entertaining language; (2) youthful deejays also aspiring to be responsible broadcasters; and (3) youthful deejays broadcasting to a youthful audience that often includes not only college students but also high school students. Together, these elements have effectively excused obscenity on college radio from the kind of scrutiny applied by the FCC to commercial radio, but at the same time have elevated the issue of "indecency" on college radio to more active FCC regulation. In fact, the FCC recently fined six college stations for indecent broadcasts, and began investigations into six others for indecency violations (*Broadcasting,* March 2, 1992).

Indecency is a subarea of obscenity, and is defined mostly by age. According to the FCC, indecency is "any offensive language that might be heard by children and teenagers between 12 and 17" (*Washington Journalism Review,* November 1990, 21). Such an inherently hard-to-define area of regulation is particularly relevant in the college radio environment because high school listeners can be exposed to an amateur deejay "slipping up" in the quest to be humorous on the air. Luckily for college broadcasters, the FCC response has generally been to forgive college radio broadcasters for obscene language as long as it occurs during the designated "safe harbor" time frame (from 9 p.m. to 6 a.m. daily) [as of 1994], when children supposedly are in bed and not listening to the radio. However, the burden of proving "no harm done" has been placed on the shoulders of college-station administrators, who must argue that the obscenity uttered on the air was an accident of a learning deejay who otherwise was certified for on-air broadcasting in a responsible training program.

Thus, college radio is a unique environment where the ideal of protecting society's teens from potentially offensive on-air language is juxtaposed with the ideal of allowing college deejays to experiment with commentary in the spirit of free expression. That environment creates a role for obscenity in college radio where a discourse I am calling "expositional obscenity" emerges from the organizational interaction of two ends of a language spectrum that conventional broadcast communication does not normally engage. At one end is the clean language authorized for on-air use. At the other end is the dirty language prohibited from on-air use. Somewhere in the middle falls the expositional obscenity of college radio.

I define expositional obscenity as an experimental form of potentially obscene language offered for public consideration by a broadcast medium at a particular moment in time. Expositional obscenity, then, is the first test of whether or not a broadcast word is to be considered by the marketplace and then officially by the FCC as obscene; the language in question can involve the use of a new word such as "muff" or the new use of an older word such as "butthead." However, expositional obscenity is but a finite stage of a much longer process of public acceptance of new words and expressions; for after the expositional stage, another stage begins where the word or phrase in question is either picked up by more conventional forms of media—like the top-forty radio station—or regulated off the air by the FCC as an adjudication of listener complaints.

Two scenarios in a college radio organization lead to the formation of expositional obscenity. I will describe these scenarios by drawing examples from the college radio stations at which I have served as deejay and/or faculty advisor.[2] The first scenario is the discourse printed in the official station manual typically used to indoctrinate new deejays into the organization. For example:

> Any member found to be broadcasting vulgarities will be suspended or expelled, depending on the severity of the offense. Vulgarities include, but are not limited to, "shit, piss, fuck, cunt, motherfucker, cocksucker, bitch, asshole, bastard, son-of-a-bitch, and goddamnit." Vulgarities also include medical terms such as "penis" and "rectum" used in a nonmedical context.[3]

A second scenario is the off-air chitchat of deejays confronted abruptly with opening up [a] microphone and saying something. To fully appreciate the skill needed for such a situation, one must consider the necessity to keep completely separate, at a moment's notice, the two extremes of the obscenity language spectrum when a song suddenly ends and a moment of dead air calls for some deejay talk. Sometimes the predicament creates humorous moments. Once I heard a deejay say, in an angry way exactly ten seconds before opening the microphone at the end of a song, "I had to wait 45 fucking minutes for the band to come on stage." Then, suddenly, when the microphone was opened, the deejay announced a calm and clean version of the event: "You just heard the Cure on 90.3 WESS, who took a lonnnnngggggg 45 minutes to get their act together last night." That example is common for youthful deejays aware of the two interacting, but separate, planes of language: "the clean" and "the dirty."

The two scenarios of obscenity in college radio I have just described—the station manual with its printed permanence and the graphic off-air chitchat with its potentially spontaneous termination—expose college deejays consistently and abruptly to the realm of the obscene. Thus, college deejays are highly aware of the seven bad words because of their ongoing anxiety that one of these words might slip out over the air if they do not pay careful enough attention to their own speech, the speech of their guests, or the lyrics of a song.

However, despite the graphic depiction of obscene words in the station manual and during off-air conversation, many words and phrases of a potentially obscene nature are spoken on college radio shows. That is because the dirty end and the clean end of the language spectrum interact in an environment where the deejays are learning yet adventurous, while the FCC is forgiving yet watchful.

The result of this interaction is expositional obscenity, the on-air leaks of potentially bad words from creative deejays anxiously seeking to shock the audience without actually using the explicit seven words. Let's take a hypothetical but very real example: As a song is finishing, two deejays open up the microphone to talk on the air. One says, "Hey, what's goin' on with Ted Danson and Whoopie Goldberg?" The other one replies, "Yeah, are they boffing?" The first one remarks, "Well, he does have a mighty big helmut when he's not wearing his toupee."

This remark introduces a common feature of expositional obscenity: questionable dialogue that engages the audience in a rhetorical way. That is to say, on-air dialogue that invites the audience to be witty enough about the implied meaning of the ambiguous words and to render a judgement about whether the language is too obscene. In the previous example, the audience is invited to judge whether "boffing" is too graphic and to figure out that "helmut" is a double entendre for penis. This exchange also illustrates that the genre of expository obscenity invites the audience to be an active participant in the meaning of the remark. Put another way: In order to figure out what the deejays are talking about, the audience must vividly imagine what the expression means in light of its context. The rhetorical nature of this process is similar to that posed by the enthymeme, where the audience fills in the missing premise to arrive at the conclusion themselves.

Such a deejay exchange would probably be ignored by the FCC unless a substantial number of listener complaints are filed with the proper documentation. And while this is certainly possible, the FCC would still be likely to excuse the exchange if apologies are offered and the deejays are disciplined internally.

The Role of College Radio in
Broadening Broadcast Language

In the meantime, however, words like "boffing" begin to gain permanence also in the mainstream language of commercial broadcast media after their repeated usage in the college radio environment. In so doing, the words or phrases of questionable obscenity move beyond the stage of exposition to the stage of general usage by the media. Such was the case for a song called "Detachable Penis," broken by college radio and later aired on commercial radio only after it was seen to be a "safe" song. Such is also the case for a song called "Asshole," an apparently obvious violation of the tabooed "seven bad words" that has not yet resulted in fines. Both songs attained their legitimacy on college radio before they were adopted by commercial radio.

Therefore, commercial broadcasters take on potentially obscene language that has often been tested first in the college radio environment. This parasitic relationship makes college radio a unique force in the evolving definition of obscene expressions. It is a force that does at times produce tasteless humor; but it is a force that on balance produces for college deejays a healthier understanding of the meaning of obscenity than either textbooks or public discussion would produce. In essence, college deejays get to witness the full process of a public conclusion that certain words are offensive in light of a sustained public reaction. Moreover, the process is played out for college deejays in much less of an ambiguous fashion than for professional broadcasters, where the potentially offensive words are not even aired until they are safe.

In closing, I believe that college radio deejays have a very real understanding of obscenity issues because they expose the new language forms to their audiences and experience firsthand whether or not the audience is offended by their remarks. Moreover, because the ambiguous nature of the expositional obscenity in question engages the mental imagery of audience, the public reaction to the language is truer to FCC-actualized definition of obscenity—an expression that some groups of people find to be extremely offensive. Therefore, college radio performs a decidedly productive role in establishing the boundaries of acceptable broadcasting expressions, and in exposing its practitioners directly to the process by which symbols become obscene.

Notes

1. They are: shit, piss, fuck, motherfucker, cocksucker, asshole and tits.

2. I currently serve as University Advisor to WESS radio at East Stroudsburg University (1375 watt output, Diversified format, faculty advised, funded through student fees). Previously I served as Faculty Advisor to KSSB radio at the California State University, San Bernardino. In addition, I have been a deejay for WIXQ radio at Millersville University and WPSU radio at the Pennsylvania State University.

3. Quoted from the Station Manual of WESS (90.3 FM).

References

Carroll, Raymond L.; Silbergleid, Michael I.; Beachum, Christopher M.; Perry, Stephen D.; Pluscht, Patrick J.; and Pescatore, Mark J. (1993), "Meanings of Radio to Teenagers in a Niche-Programming Era," *Journal of Broadcasting and Electronic Media,* Vol. 37, Num. 2, pp. 159-176.

"CPB Opponents Hoist Indecency in Funding Debate," (1992) *Broadcasting,* Mar 9, p. 36.

Donnerstein, Edward; Wilson, Barbara; and Linz, Daniel (1992), "On the Regulation of Broadcast Indecency to Protect Children, *Journal of Broadcasting and Electronic Media,* Spring, p. 111-117.

"FCC Puts Broadcasters on Notice for Indecency," (1992) *Broadcasting,* Mar 2, p. 29.

Fowler, Mark S. and Brenner, Daniel L. (1982) "A Marketplace Approach to Broadcast Regulation," *Texas Law Review*, Vol. 60, p. 207.

"High Court Agrees Ban is Unconstitutional," (1992) *Broadcasting,* Mar 9, p. 35.

"Indecency Suit Chills Campus Radio Stations," (1990) *Washington Journalism Review,* November 1990, Volume 12, Number 9, p. 21.

Pember, Don R. (1984). *Mass Media Law* (3rd edition). Dubuque: Wm. C. Brown Publishers.

Sterling, Christopher, and Head, Sydney (1990) *Broadcasting in America,* Boston: Houghton Mifflin Co.

"WLUP(AM) Hit with Second Indecency Fine," (1993) *Broadcasting,* Feb 22, p. 6.

PRACTICAL APPLICATIONS

Indecency and Obscenity

Although the issues of indecency and obscenity in programming at the college radio station have been compellingly raised in Robert McKenzie's paper, the practical applications discussed here center on the actual dealings with these issues at the station. It is my hope that station managers and advisors can learn from the experience and suggestions of others.

From the outset, as addressed in Chapter 6, the issues of indecent and obscene programming should be made clear in the station policy manual. At this point, good judgment on behalf of the station leaders is paramount. Understanding your audience (see Chapter 5) will help to determine the realm in which you operate. It is important to realize that different programming content is allowed or restrained depending upon the intended audience and time of the day (this is where the programming of indecent material to children comes into play).

The actual program idea or intent will play a role in determining how far one can go. For example, a prerecorded broadcast at the university radio station I managed dealt with the feelings of Vietnam War veterans. As one might imagine, some of the language in the program was very forceful and, by some estimates, could have been considered vulgar. Because the program dealt with the issue of war and the feelings exhibited by the veterans themselves, the language was appropriate. But was it appropriate for broadcast? I was under the impression that the student program director was going to screen the program and edit it before the broadcast. In fact, the program director let the program run as originally recorded. Because it was broadcast on Sunday afternoon, I was concerned that we might receive some listener complaints. As a precaution, I wrote a "memo to the file" detailing the intent of the program, my own awareness of the broadcast, and my understanding that the program was to be previewed for language. I believed that this would provide adequate reference if needed. It is helpful to keep in mind that each action by the FCC in regard to obscene and indecent programming is based on individual specific circumstances.

> *The most recent example that comes to mind of a noncommercial station fined for indecency was the 1993 case of the station in New York state with the unfortunate call letters WSUC. There was a nearly $24K fine for the airing of the song "Yodelling in the Valley." Trust me, the lyrics left nothing to the imagination. Of course, there have been more recent examples regarding commercial stations (Infinity, and Infinity, and Infinity ...). (NACB ListServ, March 10, 1998)*

Being aware of the station's policies and the laws governing indecency and obscenity is the responsibility of station management. The knowledge of the current "safe harbor hours" is of utmost importance for programming the station. The hours are constantly subject to change. A radio station manager's understanding of obscene and indecent material will vary in relation to his or her counterparts in other media— for example, in a cable television station operating from campus. Cable outlets are normally distributed on a local cable system and/or public access channel. These circumstances allow for more of a "free speech" environment and thus, at times, more liberal content. Quite often this kind of milieu is familiar to a student programmer on the radio broadcast. But students must understand that they are "broadcasting," a totally different medium from cable distribution (or even closed-circuit), in which other rules apply.

> In 1987, the Commission replaced its "seven dirty words" standard
> with a "generic" definition of indecency. Indecent material is now defined as:
> "language or material that, in context, depicts or describes, in terms patently
> offensive as measured by contemporary community standards for the broadcast
> medium, sexual or excretory activities or organs."

■ a. Context. The Commission has defined context only by stating that it encompasses a "host of variables" which includes the "manner" in which the material is presented, the issue of whether the offensive material is isolated or fleeting, and the "merit" of the material.

■ b. Depictions or Descriptions. The new definition extends to television broadcasts but to date no television station has been fined for broadcasting indecent material.

■ c. Patently Offensive. The standard applied is a national standard based upon what the Commission at any given time believes will offend the "average" broadcast viewer or listener. Because the standard does not look to local values or sensibilities, it is discernible primarily through rulings as to what the Commission finds offensive.

(National Association of College Broadcasters 1995, 92)

To help clarify the distinction between indecency and obscenity, attorneys at law Arter and Hadden of Washington, D.C., offer the following:

FEDERAL OBSCENITY AND INDECENCY LAW

Although the FCC may not censor programming, it has the authority to fine broadcasters for the broadcast of obscene and indecent material. Infinity Broadcasting recently agreed to pay *$1,715,000* in settlement of numerous FCC fines for indecent broadcasts on the Howard Stern Show on Infinity's stations in New York, Philadelphia and Washington.

While obscene material may *never* be broadcast, indecent material may be broadcast during a "safe harbor" period, between the hours of 10:00 p.m. and 6:00 a.m. It is thus important that broadcasters understand the distinction between *obscene* material and *indecent* material.

Obscene material may not be broadcast at any time. It is not accorded constitutional protection under the First Amendment. Material is obscene if:

(1) the average person, applying contemporary community standards, would find that the material appeals to prurient interests;

(2) the material describes or depicts sexual conduct in a patently offensive manner; and

(3) taken as a whole, the material lacks serious literary, artistic, political or scientific value.

A broadcaster who airs obscene material may be prosecuted under the U.S. Criminal Code, and if convicted, may be fined and/or imprisoned for up to two years. However, even if the broadcaster is not prosecuted for airing obscene material under the Criminal Code, the FCC has the authority to fine the broadcaster for violating the Commission's rules against obscene broadcasts.

In contrast, *indecent* material is accorded constitutional protection under the First Amendment. The FCC defines indecent material as:

language or material that, in context, depicts or describes, in terms
patently offensive as measured by contemporary community standards for
the broadcast medium, sexual or excretory activities or organs.

Although constitutionally protected, the government may regulate indecency to
promote a compelling interest if it chooses the least restrictive means necessary
to further the articulated interest. The Supreme Court has held that the FCC
may, in appropriate circumstances, place restrictions on the broadcast of indecent speech in light of the medium's "unique pervasiveness and accessibility to
children."

Over the past two decades, the FCC has prohibited the broadcast of indecent
material during those hours of the day when there is a reasonable risk that children will be in the audience. As a result, broadcasters have been required to
channel their indecent programming to a safe harbor period. Because of various
court challenges to the safe harbor period, "safe" hours have changed over the
years. Until recently, there was a safe harbor period from midnight to 6 a.m. for
commercial broadcasters and from 10:00 p.m. to 6:00 a.m. for noncommercial
broadcasters that sign off by midnight. The current safe harbor of 10:00 p.m. to
6:00 a.m. for *all* broadcasters is the result of a decision by the D.C. Court of
Appeals last summer.

In determining whether material is indecent, broadcasters should focus on
whether the material is "patently offensive." The mere discussion of "sexual or
excretory activities" will not necessarily constitute indecent speech if such
material is not presented in a "patently offensive" manner. In one case, for
example, the FCC concluded that a program addressing sexual issues was not
indecent because the material was instructional and clinical in nature.

There are risks involved in broadcasting a discussion about "sexual or excretory activities" outside the 10:00 p.m. to 6:00 a.m. safe harbor period.
Broadcasters should be prepared to justify that a particular broadcast is not
"patently offensive as measured by contemporary community standards for the
broadcast medium."

The broadcast of indecent material outside the safe harbor period is a violation of the U.S. Criminal Code as well. However, a broadcaster who airs indecent material outside the safe harbor period is more likely to be fined by the
FCC than prosecuted under the U.S. Criminal Code.

The FCC does not monitor the airwaves for obscene and indecent broadcasts. Enforcement actions in this area usually are the result of complaints from
members of the public who are offended by something they hear on the radio
or see on television.

Thus, it is important for station owners and their employees to be aware of
these federal laws against obscene and indecent broadcasts, and to know that
the only current safe harbor for indecent broadcasts is 10:00 p.m. to 6:00 a.m.

(December 1998)

Source: Arter and Hadden, LLP, Washington D.C. Used with permission.

Of course, part of the ongoing debate concerning obscene and indecent programming centers on what some believe is the vagueness of these definitions.

There truly is not a clear and decisive line between what is allowed and what would be considered objectionable. Thus, the subjective interpretation of questionable programming is quite often left to the discretion of local radio programmers, who are at times the students themselves operating the college radio station.

Good judgment and managerial control at the station will help preclude problems in the areas of indecency and obscenity. Common sense helps a great deal, too. And, as is normally the case interpreting the law, ignorance is not an excuse.

Finally, as outlined in Chapter 6, in-house training of the staff will aid in clarifying what is considered allowable programming at the college radio station. Knowledge and communication will serve to assist in the overall station operation. More information regarding rules violations pertaining to the broadcast of indecent and obscene material may be obtained from the Complaints and Political Programming Branch of the Federal Communications Commission.

Music Licensing

As this chapter deals specifically with music, it is appropriate to discuss the matter of music licensing. As a copyright issue, all stations must have clearance to broadcast recorded material. The following excerpt from the *NACB Station Handbook* gives pertinent details about music licensing:

> There may be a time when you or someone on your college station will want to use "That Song" by That Group either to broadcast directly or for background music. This is permissible, only if your station has paid the necessary licensing fees (or is covered by the school's blanket license).
>
> "Why must we pay licensing fees for the music we use?" you may ask. Licensing is divided into two areas: performances and recordings. Artists, songwriters, and publishers all receive royalties for the music they perform and record. However, money from the sales of recorded and/or sheet music is a small percentage of an artist's income compared to performance rights.
>
> When a song is played on a station, heard in an ice skating rink, bar or dance club, as background music during a TV or radio show, or even when a business phone is put on hold, that's a music performance and it's legally required for the user to have licensing rights to it.

May I See Your License?

Licensing organizations were created to solve the enormous problem of arranging the rights to every song individually and determining the royalty payments. The three current organizations which handle this task are BMI, ASCAP, and SESAC, Inc., a private, family-held company.

Each organization handles its licensing agreements differently. They each set their own licensing fees and terms of payment.

SESAC

SESAC uses the charts from various publications, such as Billboard, Cash Box, and R&R (Radio & Records). Artists are paid for making these charts, how long they stay there, and if they cross over to other music charts.

BMI

BMI relies on each station to provide logs (even college radio stations are asked to fill them out). Stations are simply asked to list continuously all songs, artists, and—of course—songwriters, aired on the station over a period of several days.

ASCAP

ASCAP tabulates the airplay of music on broadcast stations through monitoring stations. Tapes are made of stations' broadcasts secretly, then ASCAP employees identify them and the proper royalties are determined and paid out.

Licensing and College Stations

There are a ton of artists that are played on college radio and nowhere else. BMI is in the forefront of recognizing the contributions of college radio because it collects data from college radio. The implications are significant in at least two ways. First, most of the music industry sees that college stations play artists first. The licensing procedure can validate that assumption with a mathematical formula. Subsequently, the role of college stations in their minds of the managers, agents, venues, etc., will be more publicly known and appreciated.

ASCAP American Society of Composers, Authors, & Publishers One Lincoln
     Plaza New York, NY 10023 212-595-7542
BMI Broadcast Music, Inc. 320 West 57th St. New York, NY 10019
     212-586-2000
SESAC, Inc. 156 West 56th St. New York, NY 10019 212-586-3450

(National Association of College Broadcasters 1995, 166)

As already mentioned, the use of "blanket agreements" may be enforced at the school. If so, the campus radio station may or may not be included within the agreement. "If these agreements between the University and BMI, ASCAP and SESAC do not include the radio station, the agreements must be revised. And, upon such revision, the annual fees are likely to be increased" (National

Association of College Broadcasters 1995, 166). Also, although it is an issue I am not aware has ever been challenged, the use of licensed music in station promos and underwriter announcements is *not* included. These would fall under what is known as "mechanical rights," thus requiring separate licensing.

At the time of this writing, licensing for use over the Internet by stations was still being debated. Additionally, it is a good idea for the college radio station to check local franchise agreements concerning licensing and the rebroadcasting of the campus station over the local cable channel. Finally, read your licensing agreements to ensure they meet your "broadcast" needs.

Charts and Music Reps

Most stations play alternative because that's what the record labels send them free copies of the most, and in part because the students attracted to those stations like it. It also means it's easier for stations to get concert tickets, promotional merchandise, etc., from companies if they stick to that format. But the disadvantage is that the radio market in every city is saturated with alternative, and there's no way a college station is going to win a battle with a commercial station playing the same format, unless they go to an extreme niche of alternative, but then they risk alienating the campus audience they're supposed to serve. (NACB Discussion List, October 15, 1996)

A final issue concerning music is that of music publicity, including published charts (as mentioned earlier), and recommendations from music industry representatives to formulate station playlists.

One problem that arises when discussing college radio charts, playlists, and formats is the temptation to assume that formatting and programming according to the charts are bad; that they make college radio somehow less "authentic." For some, there is a rather distasteful politics associated with the sort of loosely-formatted college radio deemed by many to be authentic. (Kruse 1995, 165)

Earlier in this chapter, I mentioned that major music industry trade magazines now cover college radio playlists, along with the mainstream music. Again, Kruse comments on these publications:

There are a number of trade papers that include a variety of charts, including *Cash Box, Radio & Records,* and *Hits,* but in the world of alternative rock and pop—what I have been calling college music—the most important charts are found in *Billboard, Rockpool,* and especially *CMJ* [*CMJ* debuted in 1979 as the *College Media Journal*] and the *Gavin Report. CMJ*'s charts are the ones most watched by the industry; its figures are used in programming college radio stations and stocking retail stores. (184) [See Figure 4.1]

Magazines such as *Spin* serve college radio well and in 1998 was touted to be the "country's most widely read alternative music magazine" (see Associated Press 1998b). Also, the *CMJ New Music Monthly* plays the role of "a college radio trade magazine and new music's most reliable litmus sheet" (McDonald 1995, 20).

Playlists supply the needed information to find out what's happening in the

---

CMJ Online
new music first
Last Updated: September 8, 1998

FEATURES
artist interviews
music reviews

RESOURCES
upcoming releases
free charts
tour dates
artist links

CMJ
new music report
new music monthly
music marathon
about cmj
staff
feedback

INDUSTRY ONLY
trade side login
CMJ Playlist Reporting (BETA)

today's reviews
with RealAudio samples

 P.W. Long with Reelfoot
Push Me Again
(Touch And Go)

 Kahimi Karie
Kahimi Karie
(Minty Fresh)

musicnews
· JOB OPENING AT CMJ
· Nick Cave To Pen Bible Foreword
· Zippers Open For Bennett
· FCC Shuts Down Stations
· Depeche Mode To Tour
· DMC U.S. DJ Championship
· Play and win with CMJ Trivia

CMJ New Music Report
The leading weekly trade is available for download in PDF format. Get **e-CMJ** today.

e-CMJ

CMJ NEW MUSIC
MONTHLY
CMJ's consumer mag. This month's cover story: Rancid.

**11 Middle Neck Road, Suite 400
Great Neck, NY 11021-2301
P: 516.466.6000
F: 516.466.7159**

In March of 1979, Bobby Haber began publishing CMJ New Music Report out of his parents' basement. That very first issue of CMJ featured Elvis Costello on its cover. From that day forward, CMJ New Music Report has distinguished itself and risen to prominence as, to quote the Los Angeles *Times*, "the bible of new music."
For radio programmers and record industry executives alike, CMJ New Music Report, published weekly, is their primary source for up-to-the-minute radio airplay data, and cutting-edge new music editorial. Today, in addition to CMJ New Music Report, CMJ also publishes CMJ New Music Monthly, which brings CMJ's expertise to the music consumer -- along with a compact disc in every issue, featuring over 70 minutes of the hottest new music available; and each fall CMJ hosts the four-day music industry convention and band showcase known as CMJ New Music Marathon, MusicFest and FilmFest.

President
Robert Haber
Executive Vice Presidents
Joanne Abbot Green, Diane Turofsky
General Counsel/Chief Operating Officer
Alex Ellerson

FIGURE 4.1 CMJ Online: New Music
Source: College Media Inc. Used with permission.

music world. "Chart information [is] especially useful in promoting records to college radio; charts are a sort of shorthand record labels can use to describe a record's success or potential success to music directors" (Kruse 1995, 190).

Of course, along with the charts come the music company representatives themselves. Music reps are the direct contact to music labels that produce and distribute music to radio stations. As addressed later, in Chapter 6 ("Who's Running College Radio?"), the aspects of promotion, public relations, and pressure from outside entities, particularly record promoters, can put a great deal of strain on the college radio station programmers. The manager/faculty advisor needs to work closely with student programmers and station music directors to develop and nurture ethical relationships with record promoters. Words of wisdom need to be delivered to the student dealing with record reps to ensure that an understanding is promoted to foster a working relationship. The student must be cautious to determine the difference between recommendations and influence. (The university radio station that I managed had a policy of not accepting gifts, including meals, from record reps. Small items received in the mail unsolicited, such as desk ornaments, were acceptable. The only time that the student was allowed to socialize with record reps was at national record conventions.) As the 1995 headline so explicitly asked: "Radio Free U.: Who's running this show anyway?" (Radio Free U. 1995).

REFERENCES

Allen, S. R. (1997). College radio rocks more than the music industry. *Seton Hall University Magazine* 7(2): 13-16.

Associated Press. (1998a). Generation X rock fans gather for a Phish jam. *Denton (Texas) Record-Chronicle,* August 16.

Associated Press. (1998b). Spin gets a news spin. *Denton (Texas) Record-Chronicle,* August 16.

Avery, R. K. (1998). Educational radio. In *Historical dictionary of American radio,* ed. D. G. Godfrey and F. A. Leigh. Westport, Conn.: Greenwood Press.

Bessman, J. (1989). Swear off the 'C' word, CMJ college panel says. *Billboard* 101(46): 48, 52.

Bessman, J., and P. Stark. (1989). College radio focus of CMJ panels. *Billboard* 101(45): 12, 18.

Caton, B. (1979). *Public radio in Virginia.* Telecommunications Study Commission, Working Paper No. 12. Richmond: Virginia State Telecommunications Study Commission. ERIC, ED 183 209.

Esselman M.(1996). Angry young women. *USA Weekend,* January 12-14, 14-15.

Gimarc, G. (1994). *Punk Diary: 1970-1979.* New York: St. Martin's Press.

———. (1997). *Post Punk Diary: 1980-1982.* New York: St. Martin's Press.

Gronau, K. (1995a). NAC jazzes up ratings of age-old radio genre. *Radio World Magazine* 2(10): 32, 34.

———. (1995b). New artists, public broadcasting, NAC regenerate traditional jazz. *Radio World Magazine* 2(11): 20, 22.

Gundersen, E. (1989). College radio explores rock's flip side. *USA TODAY,* February 27.

Holterman, S. (1992). The relationship between record companies and college music directors: A descriptive study of alternative radio. Master's thesis, University of Tennessee, Knoxville.

Knopper, S. (1994). College radio suffers growing pains. *Billboard* 106(28): 84.

Kruse, H. C. (1995). Marginal formations and the production of culture: The case of college music. Ph.D. diss., University of Illinois, Urbana, 1995. Abstract in *Dissertation Abstracts International,* 56/09:3360.

Marcus, T. (1997). Tuning in to college radio. *Link, The College Magazine,* October/November, 26-27.

Mayhem, M. (1994). A little college music, please. *Fort Worth Star Telegram,* December 23.

McDonald, G. (1995). Left of the dial. *U. Magazine,* April 20-21.

McKenzie, R. (1994). The role of obscenity in college radio. Paper presented at the Freedom of Expression Division of the 1993 Speech Communication Association Convention in Miami, Florida, and the 1993 Speech Communication Association Convention in New Orleans.

Mundy, C. (1993). Radical radio. *Mademoiselle,* November, 70.

National Association of College Broadcasters. (1995). *1995 NACB station handbook.* Providence, R.I.: National Association of College Broadcasters.

Note Book. (1998). *Chronicle of Higher Education* 54(23): 13, A52.

October sample. (1994). *M Street Journal.* New York: M Street Corporation.

Ozier, L. W. (1978). University broadcast licensees: Rx for progress. *Public Telecommunications Review* 6(5): 33-39.

Pareles, J. (1987). College radio, new outlet for the newest music. *New York Times,* December 29.

Pitts, M. B. (1995). The week for starters: Zeniths and nadirs. *USA Weekend,* June 9-11.

Protect your freedom. (1996). *Radio World* 20(3): 5.

Radio Free U. (1995.) *U. Magazine,* April, cover.

Ryan, L. N. (1997*). Highlights of the public radio programming study, fiscal year 1996.* CPB Research Notes No. 105. Washington, D.C.: Corporation for Public Broadcasting.

Sauls, S. J. (1995). College radio. Entry submitted to the *Encyclopedia of United States Popular Culture.* Santa Barbara, Calif.: ABC-Clio, forthcoming.

———. (1998a). The role of alternative programming in college radio. Paper presented at the Annual Meeting of the Southwest/Text Popular Culture Association/American Culture Association, January 30, 1995, Lubbock, Texas. ERIC, ED 416 529.

———. (1998b). The role of alternative programming in college radio. *Studies in Popular Culture* 21(1): 73-81.

Schoemer, K. (1992a). The art behind Nirvana's ascent to the top. *New York Times,* January 26.

———. (1992b). Some alternative boundaries fall. *New York Times,* October 30.

Stark, P. (1993). CMJ examines the effects of success on alternative music. *Billboard* 105(47): 13, 90.

Starr, V. (1991). CMJ aims to broaden scope of alternative scene. *Billboard* 103(44): 30, 32.

Stearns, D. P. (1986). Big bands on campus: A look at future hits. *USA TODAY,* October 24.

Thompsen, P. A. (1992). Enhancing the electronic sandbox: A plan for improving the educational value of student-operated radio stations. *Feedback* 33(1): cover, 12-15.

The usual suspects. (1998). *Unlimited Action, Adventure, Good Times,* Winter, 44, 46.

Ward, E. (1988). Back to school cool. *Mother Jones* 13(7): 47.

Wilkinson, Jeffrey S. (1994). College radio: Farm team or free form funhouse? *Feedback* 35(1): 4-7.

Wolper, A. (1990). Indecency suit chills campus stations. *Washington Journalism Review* 12(9): 54.

Zimmerman, K. (1989). Schizoid college rock: Hip radio, safe acts. *Variety,* August 16, 67.

———. (1992). Alternative pops pop's balloon. *Variety,* March 9, 64, 66.

5

Who's Listening to College Radio?

Because the funding of college and university radio stations is usually limited (as discussed in Chapter 7), many stations also solicit program underwriting support, listener contributions, and outright donations, which are important because advertiser/commercial content is severely restricted by law on noncommercial stations (Sauls 1995a). Quite often those supporting the station, either on or off campus, desire some idea of the number of listeners of the station. Of course, this is also a natural concern within the station itself: "Who's listening to our station?" and thus, "Who is our audience?" (see Steinke 1995). As in commercial radio, "with deregulation and the ownership of a majority of large-market stations by major broadcasting conglomerates, much more money is at stake now and there is a lower tolerance for unproven programming styles. So how do stations know they are on the right track?" (Hyde 1997, 26).

Being "right on track" can mean, when it comes to programming, that the ultimate consideration must be your intended audience. In the college radio station realm, that audience quite often comprises two groups: the student body and the community at large. (See Chapter 8 for a further view of communities both on and off campus.) Determining who makes up these listening communities is an ongoing challenge.

SURVEY RESEARCH AND COLLEGE RADIO

Although many campus radio stations try to copy or emulate their commercial counterparts, determining who's listening to the station is often a difficult thing to do. The reality is that most college radio stations cannot afford the commercially available "rating services," such as Arbitron. Therefore, it is common for stations to undertake in-house listener surveys. Quite often these surveys are conducted by classes studying mass communications, survey design, or audience research. Although purely scientific research methods are not always employed, the results do at least provide some type of "reading" on how the campus or community at large views the station (see Sauls 1997).

Additionally, quite often the eclectic/alternative formats presented by college and university stations are solely directed at their student body, thus making it hard to identify the oftentimes "mobile" audience. Still, though, inherent to the particular needs of campus radio stations (such as pleasing the sponsoring and/or funding group of the station [Knopper 1994, 84]), the station is sometimes under great pressure to determine its current and potential audience. (For discussion purposes, please see the Practical Applications at the end of this chapter. Besides focusing on a national survey concerning audience research availability at college radio stations, general observations concerning survey research are presented. Additionally, the reader is directed to Appendix B (see Sauls 1979), where a discussion of findings from a study conducted by the author as a graduate student, highlighting specific outcomes of an audience survey, is presented.)

DEFINITION OF TERMS

Apart from the basic terms for ratings, which will be addressed later, it is first necessary to distinguish between terms used in survey research. These will come into play more specifically when a station is considering not only the number of listeners, but also such factors as program evaluation, consistency, and validity.

Wood and Wylie, in *Educational Telecommunications (1977),* discuss the subjects of research, evaluation, and validation. From the outset the authors make clear that the differences between these terms cannot be cleanly made, and the processes are not that distinct—indeed, their definitions make it obvious that there is considerable overlap. The authors helpfully note that "less formally we can define research, evaluation, and validation by the sorts of questions that we tend to ask when we are engaged in each of these" (312).

> *Research:* The systematic process of objective study and investigation of a problem, using the scientific method and inductive reasoning, in order to establish facts or truths. (312)

Here we speak of the actual process of discovering information about something. For our needs that may include who's listening and how often they are listening. In research, investigation falls into three basic categories: (1) Historical research: Why did this occur, what happened, and what were the causes and effects (both on the past and present) of this particular event? (2) Descriptive or normative research: How many of this and that are there, is this related to that, and what is the actual situation (or norm)? (3) Experimental research: Does this cause that, and can we state any principles or conclusions that add to our broad understanding of the phenomenon? (312-13)

> *Evaluation:* The process of analysis and examination of a specific project, activity, or event, in order to determine future action. (313)

Thus, the distinction is made that evaluation, as compared to research, usually is concerned with appraisal of a specific activity, program, or some other single identifiable thing. For us in broadcasting, this is an important concept in that determination of a future course of action is critical in regard to the end result of any evaluation process. As a result of evaluation we can ask such questions as (1) Do we continue the program without change? (2) Do we modify the program? or (3) Do we cancel the program? In short, are we accomplishing our goals, or do we need to change? (313)

> *Validation:* The process of determining if predetermined objectives are consistently achieved when a specific instructional unit, a lesson or course [or program], is presented to a target group of [subjects/listeners/students] (who meet certain standards). (313)

Here we become concerned with consistency. Basic survey theory purports that two aspects must always be present: reliability and validity. In order to be truly valid, an ideal must first be reliable. In other words, you can have reliability (consistency) without validation (it can be consistently incorrect), but you cannot have true validity without reliability (it cannot be valid if it is consistently wrong).

RATINGS TERMS

In the November 1995 issue of *Radio World Magazine,* Mike Burnett provided some of the most commonly used/relevant ratings and advertising terms applied to radio (44). A few of those terms are presented here, specific to use in noncommercial radio:

- *AQH*—Average Quarter Hour
- *Book*—A rating period. Typically, this covers 12 weeks.
- *Cell*—The narrowest demographic bracket. The standard age brackets are 12-17, 18-24, 25-34, 45-54, 55-64 and 65+.
- *Cume*—The number of people who tune into a station during a daypart [particularly morning and afternoon drive times, plus evenings for student-oriented music].
- *Daypart*—A definition of time in a ratings book (or rate card) that includes time of day and day(s) of the week.

■ *Demographic*—A particular age and gender bracket. [Here the research of a particular group, such as college students, is considered.]
■ *Demographics*—The age and gender characteristics of a group of people. Note the difference in usage between singular and plural.
■ *Frequency*—Repetition. How many times the average listener will hear something.
■ *Rating*—An estimate of the size of a station's audience expressed as a percentage of the total population of the area being measured. Can be applied to cume or AQH.
■ *Share*—An estimate of the size of a station's audience expressed as a percentage of the people listening to radio during the daypart.

In addition, marketing reports from such services as Arbitron ratings contain descriptions of methodologies, selected glossaries, and instructions in the utilization of findings (see Arbitron Company 1988a, 1988b, 1988c). I highly recommend that students become familiar with these sources to enhance their understanding and clarification of radio ratings and audience estimation in general.

PRACTICAL APPLICATIONS

Audience Research Availability

Most college radio stations simply cannot afford the commercially available "rating services"; however, you should be aware that for noncommercial stations "reduced" rates for radio market data are available from Arbitron. The Radio Research Consortium in Silver Spring, Maryland, can provide market analysis, including audience distribution, cume sharing, and program estimate detail. This data can be used for comparisons between noncommercial stations in the given market, as well as for analysis with the market and commercial station data (Radio Research Consortium, letter to KNTU-FM, University of North Texas, Denton, February 12, 1996). Other research services are available from Shane Media Services, Paragon Research, Rating Point Management, and Research Director (Hyde 1997, 28) and Strategic Media Research (see Fig. 5.1).

At the beginning of this chapter I mentioned that many college and university radio stations perform in-house listener surveys, often conducted by classes studying audience research. To help further understand trends in survey research at college and university radio stations, the findings contained in the *1995 College Radio Survey,* conducted by the National Association of College Broadcasters, highlight significant points. The survey results suggest several considerations for those who deal with audience research and revealed several important characteristics about listening trends of college radio stations.

Most of the stations (65.8 percent) responding to the survey had conducted some type of audience research. Among those who did, 64.0 percent had con-

Radio's ultimate audience tracking system!

For the first time ever, you can make programming, marketing, and management decisions based on up-to-the-minute facts!

Exclusively from **Strategic Media Research**, AccuTrack is a valuable new tool for programmers and managers that consists of three key components:

(1) Audience tracking: Weekly updates delivered electronically give you advance information on listening behavior in your market -- allowing you to see trends weeks or even months in advance.

(2) Marketing tracking: AccuTrack will tell you which of your advertising and marketing expenditures are working -- with **weekly** updates when your marketing is in the field.

(3) Perceptual tracking: AccuTrack will allow you to track key station perceptions year-round -- on both yourself **and** your key competitors.

Get information even faster electronically with our NEW Windows based AccuTrack software.

FIGURE 5.1 Strategic Media Research - AccuTrack
   Source: Strategic Media Research. Used with permission.

ducted a station survey, and 34.7 percent had conducted Arbitron (or other industry standard) research. Twenty-seven percent of the respondents who conducted research reported that they had conducted a nonstation survey. Most of the respondents described their research as either scientific (16.2 percent) or careful and accurate (42.3 percent), but 41.5 percent described it as informal (see Fig. 5.2).

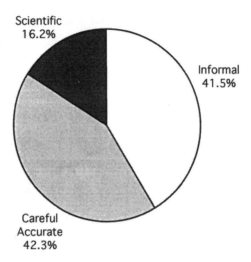

FIGURE 5.2 Perceived Level of Research Quality (*n* = 130)
> Source: *1995 College Radio Survey,* National Association of College Broadcasters, p. 26.
> Used with permission.

Additionally, there was considerable interest in conducting more research in the future. Forty-seven percent of the respondents were considering conducting a survey in the future, and 20.8 percent were giving future surveys top priority. Further analysis of the data showed that, based upon their best audience research available, 21.6 percent of the respondents estimated that they had a potential audience of 300,000 to 1 million, and 13.2 percent estimated a potential audience of over 1 million.

Thirty-seven percent of the respondents reported an institutional enrollment of 2,500 students or less, and 31.8 percent reported enrollment of more than 10,000 students. The smaller schools had a larger percentage of students in residence.

Average estimates showed that the number of students tuning in at least once a week ranged from 36.7 percent to 49.2 percent. Most respondents reported that the station was available to over 85 percent of their students. Schools with enrollment of 2,501 to 5,000 reported the lowest percentage of students who tuned into the station at least once a week (36.7 percent). Schools with enrollments of 2,500 or less reported the highest percentage of students who tuned in at least once a week (49.2 percent) (see Table 5.1 and National Association of College Broadcasters 1995, iv-v, 46-47).

Recommendations for Future Studies

The following recommendations are put forth for use and consideration by those interested in conducting future studies at college and university radio sta-

TABLE 5.1   Student Listening Rates

Number of Students*	Percentage of Students Receiving Station	Percentage Listening at Least Once a Week
2,500 or less (n = 57/49)	95.4%	49.2%
2,501 to 5,000 (n = 26/20)	89.2	36.7
5,001 to 10,000 (n = 23/13)	88.8	41.2
10,001 to 20,000 (n = 30/19)	86.1	37.6
Over 20,000 (n = 20/14)	88.0	39.6

*Percentages in each cell of Column 2 were calculated by averaging the number of students receiving the station divided by the total enrollment of each school. Percentages in each cell of Column 3 were calculated by averaging the number of students tuning in at least once a week divided by the total enrollment.

Source: *1995 College Radio Survey*, National Association of College Broadcasters, p. 30. Used with permission.

tions. These recommendations are based on ideas that I originally published in the *1995 College Radio Survey*.

First, conduct a pilot study of the survey; then provide preliminary results to the client, adjust the survey as needed, conduct an additional pilot study if needed, and, finally, conduct the actual survey. The purposes of the pilot study are to identify and confirm selected factors inherent in the study itself. Statistical testing of the pilot study responses could be performed in order to analyze the significance of illustrative findings and to establish variable reliability.

Without a pilot study, it is not possible to guarantee that the results provided by the actual survey will be those wanted by the client. In other words, without providing preliminary results, it is difficult to determine if the survey is providing the information needed. To this end a schedule might be constructed as follows:

Client Consultation
Survey Instrument Preparation
Pilot Study Preparation
Pilot Study Conducted
Preliminary Results Provided to Client
Additional Pilot Study (if needed)
Finalization of Actual Survey Instrument
Mail Out/Phone Survey Conducted
Mail/Telephone Follow-up
Final Telephone Follow-up
Survey Complete

Surveys Coded
Data Entry
Data Analysis
Report Development
Report Delivered to Client
Conference Focus Group

Review the survey dates to ensure that semester (school) dates are appropriate for the survey population. Surveys involving colleges and universities show that timing is crucial. In order to generate the greatest possible number of completed surveys, it is critical that survey dates coincide with school dates. Ideally, surveys should be conducted within the time period of September-November or February-April (summer dates are unpredictable and not recommended.)

Survey a prorated sample within the population. Statistically, it is not required to sample the entire population. In the selection of the sample, an appropriate size for a simple random sample (n) needs to be determined for the population (N). The appropriate sample size for the survey is derived using a population and a sample size for permissible error (proportion) and a predefined confidence level (typically 90, 95, 98, and 99 percent) (McCall 1982, 330).

Normally, an initial return rate is projected from the sample size. In order to compensate for loss and to increase the sample size proportionately, an adjusted sample size for the expected rate of response (na) could be calculated utilizing the preliminary estimate of the sample size (n) and an expected rate of response expressed as a proportion (Pr). Thus, utilizing the formula $na = n/Pr$ (McCall 1982, 205), an increase in the sample size to adjust for nonresponses (underestimate) could be available (na) and provide the actual number of initial surveys.

The simple random sample (na) could then be selected by numbering all potential sample analysis units (N) and conducting a random sample without replacement utilizing a computer-generated set of random digits. Random digits would be selected until the sample needed (na) was obtained.

Finally, if the study is commissioned by an outside organization (possibly the student government association), consider using corporate letterhead in mailed surveys. This can provide a source of increased returns as the respondent is communicating with a well-established organization. Caution should be used in this technique, as it could appear that the client commissioned the study with intended results. Additionally, any biases toward the client or the organization always need to be considered in undertaking any survey (Sauls 1995b, 49-52).

A Few Final Thoughts

Concerning ratings in general, Dain Schult provided some insight into the perceptions of a station's performance when he wrote in 1995: "Of course, the argument for having tangible yardsticks with which to measure your station's performance certainly has merit. Everyone needs a hopefully objective report card to gauge performance and listener satisfaction. However, when ratings become the end-all, be-all of a station's existence, the question then becomes, 'Is the tail wagging the dog?'" (43). Hence, as college and university radio stations employ in-house listener surveys and/or utilize commercial rating services, they should also consider the impact that such results could have on programming.

As a manager of a college radio station, one might question whether or not to share the findings of listener research with the student staff members. Quite often students working at the station have the idea that their audience is enormous, whereas, in reality, ratings will confirm just the opposite and "burst their bubble." It may not be a bad idea to retain findings for managerial use (such as during budget meetings with school administrators) and for applications to basic programming decisions.

If you are surveying your student audience, remember that college populations are cyclical in that students graduate every year while new students arrive annually. This provides for a continual turnover in the campus population, and thus listening habits will naturally change with these shifts.

Finally, determine who your intended audience is, then survey a sample of that segment of the population. Is the station programming directed to students only, the campus community (faculty, staff, and students), or the local community at large? In other words, is your station reaching those whom you are targeting? Don't forget that the goal of research is to provide a realistic reading of your station's audience (see Sauls 1996).

REFERENCES

Arbitron Company. (1998a). Description of methodology. *Radio Market Report: Dallas-Ft. Worth,* M3-M6.

———. (1998b). Instructions for estimating the reliability of audience rating estimates. *Radio Market Report: Dallas-Ft. Worth,* M1.

———. (1998c). Selected Arbitron terms. *Radio Market Report: Dallas-Ft. Worth,* M7.

Burnett, M. (1995). Deciphering the terms. *Radio World Magazine* 2(11): 44.

Hyde, D. (1997). Little room for error: With big bucks at stake and less flexibility in programming, radio relies on research. *Tuned In,* May, 26, 28.

Knopper, S. (1994). College radio suffers growing pains. *Billboard* 106(28), 84.

McCall, C. H., Jr. (1982). *Sampling statistics handbook for research.* Ames: Iowa State University Press.

National Association of College Broadcasters. (1995). *1995 college radio survey.* Providence, R.I.: National Association of College Broadcasters.

Sauls, S. J. (1979). *A study of radio listening habits of the residents of Denton, Texas.* Denton, Tex.: KNTU-FM.

————. (1995a). *College radio.* Paper presented at the 1995 Popular Culture Association/American Culture Association National Conference, April 14, 1995, Philadelphia. ERIC, ED 385 885.

————. (1995b). Recommendations for future studies. In *1995 College Radio Survey.* Providence, R.I.: National Association of College Broadcasters.

————. (1996). Practical applications of survey research at college and university radio stations. Paper presented at the Fall 1996 Student/Faculty Conference of the Texas Association of Broadcast Educators, San Antonio, September 6, 1996. ERIC, ED 410 626.

————. (1997). Survey research and college radio. *Feedback* 38(1): 10-12.

Schult, D. (1995). Are ratings overrated? *Radio World Magazine* 2(11): 43-44.

Steinke, G. L. (1995). Is it possible to create a significant audience for your campus station? *Feedback* 36 (2): 24-26.

Wood, D. N., and D. G. Wylie. (1977). *Educational telecommunications.* Belmont, Calif.: Wadsworth.

6

Who's Running College Radio?

*Amazing thing about students—give them an inch and they take 10 feet. ...
Needless to say, but I am not a very popular person right now. ... I see this
decision improving the station in both the short and long term, and I am letting
that vision guide my choices. No one likes changes, but they have to happen.
Otherwise we'd remain stagnant. (Personal communication to the author,
October 25, 1995)*

STRUCTURE: MANAGEMENT AND ADMINISTRATION

Although management styles and theories differ among administrators
of college radio stations, some views concerning college radio, both
on and off campus, tend to be consistent. The day-to-day operation of
such stations is totally dependent upon the ideals undertaken by managers and
put into effect at the station itself.

To further understand the concept of college radio, I focus in this chapter
upon the structure and management of college radio stations. In particular, the
control of the operation of the station itself is analyzed. Included is a detailed
study of the inherent components of such facilities, including ownership.
Additional topics include operating within an academic environment, student
staffing and training, the role of the faculty advisor, and the allegiance between
funding and control (see Sauls 1997).

COMPONENTS AND OWNERSHIP

Wood and Wylie indicate in their book *Educational Telecommunications* that
"certain common elements are present in varying degrees" in every noncommercial radio station. Here the authors are specifically addressing what they
call the *Twelve Components of a Public Broadcasting Station,* which could be

used as guidelines "to evaluate the structure and performance of any local public station," including the college radio station (72).

For application to college radio, all 12 of the components are legitimate and a good overview of the structure to be found in such an operation: 1) public broadcasting philosophy, 2) community needs, 3) governing board, 4) management and staff, 5) financing, 6) community involvement, 7) FCC license, 8) physical plant, 9) nonlocal programming, 10) school relations, 11) publicity and promotion, and 12) audience feedback (Wood and Wylie 1977, 72-75). (Theoretically, all college and university broadcasting is "public broadcasting," except for those few stations commercially licensed.) These components provide a good vehicle in which to analyze a particular station and its structure. It is therefore a good starting point to determine how the station is operated ... and thus, who is running the station.

The authors also purport that when studying the structure and operation of a station it is helpful to consider the categories of station ownership, which they break down into four traditional designations: 1) university stations, 2) school-owned stations (those owned by school districts, for example), 3) state-owned stations, and 4) community stations (75-78). Of course, the designation applicable to college radio is the category of university/college-owned stations, of which the authors note that academic resources (i.e., faculty) can be drawn upon. Station management, however, typically recognizes more of a need to program toward the general "outside" community as a whole (76).

Within the context of the campus, the college radio station can be operated under the direction of an academic department (as discussed in the next section), by direct school administrative control, by the student government, as a "radio club," as an NPR affiliate (basically professionally staffed), or as an on-campus independent operation. Whichever style is used, factors such as programming responsibilities, day-to-day operations, community service (both on and off campus), and student involvement all come into play.

Also look at who holds your license! Our license was held by the board of regents, so airing religious programming became a question of separation of church and state! We finally had to make the rule that you could play religious music but couldn't promote or condemn [any] religion, its beliefs or practices. The music, just the music, and associated commentary (that was, Band X from someplace with their newest CD). (NACB Discussion List ListServ, September 20, 1996)

THE ACADEMIC ENVIRONMENT

Our biggest challenge has been trying to convince our faculty/staff what our mission is, and why we should be considered a valuable member of the liberal arts education. When I see/hear a college station that does not have a defined format, and consistently pushes the envelope with its programming, it concerns me. (NACB ListServ, September 29, 1997)

As mentioned earlier, normally college radio stations are housed within an academic department of a school or college dealing with some form of communications studies (radio/television/film, journalism, mass media/mass communications, speech, and so on). Thus, the station is operating under the auspices and direction of an academic province. In other words, a business-type day-to-day operation (the radio station) is running in a nonvocational world (academics). And so, as might be expected, problems do arise. Format needs and wants will vary in the station, within the academic department in which it is housed, across campus, and within the community at large.

From the outset members of the academic department will, at times, voice their opinion as to the operation of the station. Although faculty members make suggestions about the operation and programming of the station, they do not actually participate in the operation. These recommendations on how to "run the station" from those not directly involved, though made under the best intentions, are not always welcomed by station management (who also is usually at least part of the department—either faculty or staff).

The manager or faculty advisor quite often views these suggestions as negative criticism. Additionally, many times the manager/advisor sees [him- or herself] as the only one participating from the department in the operation of the station. So, the manager/advisor's internal response to such recommendations is "if you want to change the station, why don't you help me!" (Sauls 1996, 21)

THE FACULTY ADVISOR

Quite often college radio stations are supervised by a faculty advisor who fulfills the duty of station manager. (Sometimes this position is a full-time staff position.) This individual oversees the administration and operation of the station on a day-to-day basis, providing the needed continuity as the student staff changes year to year (see Sauls 1995). Thompsen, in 1992, wrote that "a faculty advisor can be a driving force in shaping a vision for the station, the reasons for its existence" (14). Such demands of station administration and supervision are often in addition to the faculty advisor's normal workload requirements of teaching, research, and service (see Sauls 1996, 21). So it is not odd for the individual manager, at times, to place the operation of the station in a secondary position. The fact remains that the faculty advisor is a faculty member first and a station manager second. But the sole responsibility for operating the station lies with the faculty advisor. When problems arise, it is his or her duty to handle the situation. And these problems reflect on the faculty advisor's ability to manage the station.

The faculty advisor is the person who deals with students and station matters. Case in point revolves around CD and record theft. Here the solutions run from labeling the material beyond belief, to using "dummy" or fake disks, to installing fake security cameras in the studios, to purchasing lockable cabinets with key distribution systems.

> *Sony makes a CD Jukebox system that could be a good replacement. The system includes a standard single disk CD player and a separate 100 disk changer that can be locked away somewhere. ... You can also lock out certain tracks (say, the naughty ones) and create groups of tracks and CD's (say, for overnight play?). (NACB Discussion List ListServ, April 30, 1996)*

Of course, there are the strange and sometimes funny duties confronting the faculty advisor:

> *We have a problem with our tower. It has become home to around 150 wasps. We have considered sending someone up there with wasp spray, but it is unlikely they would get back down without being stung severely. We have urgent work that must be done ASAP, but can't do anything until we get rid of the wasps. The tower is only 180' tall, and we have considered trying a pressure washer to hose off the nests, but I don't think it will keep them from coming immediately back. Has anyone ever run into this problem and successfully solved it, and if not any untested ideas would also be appreciated. (NACB ListServ, June 15, 1998)*

One aspect that may assist the station manager and the station in general is the realization that quite often the audience itself (including the school administration, other departmental faculty, and the student body) does not understand what the station is trying to accomplish. Thus, there is "a need for the station and individual staff member[s] to 'educate' the audience on what college radio and what [their] particular station is all about" (Reese 1996, 19).

As the roles and responsibilities of those "in charge" of advising college radio stations grow, programming the electronic media entity will become even more important. Because the station is licensed to the college or university, the school, as licensee, is responsible for all that's broadcast on the station. Even if the station is a nonlicensed operation (for example, cable FM), the school is still responsible for its programming. The leadership, guidance, and capabilities of the station advisor will foster the efforts of the student staff to work within these responsibilities.

STAFFING

The number of people needed to run a full-time (or even part-time) college radio station—by generally part-time, mostly volunteer staff—can be both staggering and impressive. (The most I have ever heard of was around 170. But

50 to 70 is not uncommon.) These stations are staffed by volunteer "nonprofessional" students, along with possible skeleton part- or full-time paid staff member(s). Basically, these are full-time entities operated by part-timers and volunteers. (Station staff position responsibilities are outlined within the Practical Applications at the end of this chapter.)

To further the dilemma, natural academic attrition means that a quarter of one's staff will graduate every year. Of course, this provides college radio with one of its unique features—all stations are different, and each is constantly changing. However, the challenge remains for the manager to deal constantly with a changing staff. The one thing you can count on to remain the same is that the college radio station will constantly be changing!

> *Here at ____ State, we pay our morning and afternoon-drive guys and our overnight guy. We do so because we want consistency on those shifts and they work five days a week. We also pay our music director, asst. MD, PA Director, asst. PA director, and have work-study folks manning the phones and reception. Our GM and PD are professionals. (NACB ListServ, September 26, 1997)*

STUDENT TRAINING

"There's one thing everyone in college radio agrees upon—that the primary function of the campus radio station is to educate and train students to enter the forbidding realm of professional broadcasting" (McDonald 1995, 21). "College radio is truly the training ground for tomorrow's broadcasters, providing the student an opportunity to practice techniques in broadcasting" (Sauls 1995). Of course, part of this training is to allow for the actual running of the station. Like internships at commercial stations, the college radio station attempts to give students the opportunities to work in a somewhat professional environment, make mistakes, and learn from their experiences (see Ryan and Baruth 1993).

The point about students running the station is an important one. "Here it must be recognized that 'because of their limited life experience, students may not always know the difference between promotion, public relations, and pressure (Holterman 1992)' from outside entities, particularly record promoters (Wilkinson 1994)" (Sauls 1996, 20). (See Chapter 4 for a discussion of the record-promoter issue.) Although the question of what role college radio plays in training students (see Kruse 1995, 179-180), and the perceptions of college radio in general, are debatable (see Reese 1996 and Sauls 1993), it is here that station administration and management play a most critical part.

FUNDING AND CONTROL

The necessary questions must be asked: Who pays for college radio? And who controls college radio? (Funding of the college radio station will be discussed further in Chapter 7.) The fact of the matter is that many college radio stations receive their funding via some arm of the student government, which usually is in charge of overseeing (or at least recommending) the allocation of student fees. Furthermore, those of us in college radio are constantly battling with the idea that though students are paying for the support and operation of the station through fees, their input is limited as to what the station actually programs. Thus, controlling the operation of the station to ensure that the students are content with its sound is a viable concern. (Note: Most likely these stations are under the total control of the funding source, such as the student government or direct school administration.)

Because most campus radio stations are under the auspices of an academic department within the college or university, the presence of the station can complement actual course work. The station faculty manager/advisor here plays the ever-important role of connection between the cocurricular activities of the station, and departmental courses and academics. Because of this relationship between the station advisor and academic department, "financial resources and operating procedures are almost entirely determined by the academic missions of the department" (Ozier 1978, 34). The findings of various studies reflect the important association between academic programs and the funding and purposes of college radio (see Sauls 1993 and Sauls 1996). But is the department running the college radio station?

Finally, the question of control and funding becomes particularly pressing when the station coverage of school events, such as athletics, begins to take precedence over normal programming and station operation. Here, as a college radio station manager, one may begin to question the supposed allegiance to student staff, station audience, and the funding source. The situation may even deteriorate to the point that outside entities (other stations in the market) begin to replace the "student" operation—for example, commercial radio broadcasters may wind up participating, via simulcasting, in school athletic events. Conflicts can arise when the students who operate the radio station believe that, because the station is student funded and student operated, the broadcasting of campus sports should be undertaken by students. Thus, the true question does indeed become, "Who's running college radio?"

RECENT STUDIES ON COLLEGE RADIO ADMINISTRATION

In this section are listed numerous academic studies on college radio station administration and management. They are sources for further investigation in this area. (Full citations for the works may be found in the reference list at the end of this chapter.)

A recent study of note was that of C. E. Hamilton entitled *The Interaction between Selected Public Radio Stations and Their Communities: A Study of Station Missions, Audiences, Programming and Funding* (1994). Although the study did address the notion of funding, the survey was limited to eight public radio stations, to establish the idea of two independent continua (community of service and pursuit of audience) by which public radio stations may be differentiated. Another 1994 study researched the administrative patterns of on-campus radio stations in regard to the leadership behaviors of managers. This study (Dennison 1994) was limited to NPR affiliates only. Some studies concerning college and university radio have presented a thorough investigation and description of the state of noncommercial radio in higher education (Leidman 1985). Approaches have also been taken to outline specific theories within college and university radio (Poole 1989).

Kruse's 1995 dissertation (mentioned earlier), while addressing the issue of college music, also delves into the aspects of the relationship between working at a college radio and the training of future broadcasters—a management perspective within an academic entity. She quotes a music director at the University of Illinois's commercial, student-run station, WPGU: "You've got the door open for you because you work at a real station. If I'm music director at PGU, it's almost the same as being music director at any other station of the same size. And that's the great thing, because it's so easy to get a position at the station, and it gives you such an array of choices in the industry, it'll get you in that door" (Kruse 1995, 180-83).

Finally, Bailey's 1993 study deals specifically with the perceptions of professional radio station managers in regard to the training and experience that potential employees receive at college radio stations. Of interest, the work details extensively the ideals of instructional broadcasting (Bailey 1993, 62-99). His work also examines administrative patterns and skill areas of potential employees.

Many of these works are cited throughout this book. Of course, all of these studies contain numerous references that can also be researched.

PRACTICAL APPLICATIONS

Management: As a Manager for and with Personnel

Whether the school radio station is a new entity or one that is 25 years old, the day-to-day administration and oversight of the station operation is under auspices of the station manager. This individual will be working with employees (either current or new) who are the working personnel of the station—most likely students. Even more interesting, quite often the manager is looking for those who will commit a great deal of time to the station.

> New members must be welcomed into the station and made to feel important. Giving a specific job title and duties will make staff members feel that they are an integral part of the station. With these responsibilities must come welcomed input. How a situation is approached can make a world of difference: being told

to do something and how to do it is far different from being addressed with a problem and asked for input. The key is open communication. The more staff is treated like responsible, thinking individuals, the more motivated the staff will be to act responsibly and think independently. Another tip: never ask the staff to do something that you wouldn't do yourself. (National Association of College Broadcasters 1995, 16)

From the outset, this involves plans, objectives, goals, praise and criticism, personalities, and so on. All of these elements are addressed continually by a good manager (especially organizational skills), and, more important, a good personnel manager. Remember that people are the heart of any organization.

As is discussed in Chapter 10, letting your staff know that they are professionals who hold responsibilities will make a world of difference. Quite often this entails letting them participate in station decisions. It's not that they will be doing your job as station manager, but that they are truly participating in the operation of the station—even the goals and results that will take hold long after they graduate and leave the station.

Training Programs at the College Radio Station

Try your best to impress upon the staff that they are continuing work that was begun long before they arrived, and that they are providing the continuity so important for an ongoing operation such as is the radio station. This is where the training program comes in. "With a solid training system [the] staff will feel better about their performance and actually sound better. They will have a greater appreciation of the station overall, not just of their department. That will also result in higher morale, which makes accomplishing every station project easier" (National Association of College Broadcasting 1995, 19).

Training involves everything you can imagine. The extent of the in-house training program at the station is left to the discretion of the station manager and station staff. Everything from an overview of general station policies to the specifics of how to answer the station telephone can be included in the training regime. This should be a program that is ongoing and ever-evolving. Not only will a good training program help in the day-to-day operation of the station, it is critical during periods of crisis. (I survived a crisis of my own wherein the campus radio station that I managed was completely devastated by a fire. That's when you really need help from the station staff, in addition to the outside commercial world. Knowledge and ability by the station personnel really paid off!)

Though an effective training program will make a big difference in the way a station operates, don't expect to produce miracles. There are certain things that cannot be taught: dependability, the ability to accept and act on constructive criticism, thinking before speaking, the skills of listening and reflecting, and asking questions. A key component for any staffer is learning where one can be flexible. For example, at what point is a technical problem serious enough to call the Chief Engineer at home? (National Association of College Broadcasters 1995, 22)

A final note concerning training provides for exactly how detailed the "system" can be. Components can include setting, timing (by week, month, semester, etc.), scheduled meetings, interviews and selection, applications (on-the-job, practical, staged, etc.), studio on-air and production procedures, examination and testing (written, oral, and procedural), and rewards (i.e., end-of-training party).

Management Perspective on Station Legalities

The complainant called the station, and then called the college president. He told her to complain to the FCC, and she did. The station received a $23,700 fine. (DJ-L Campus Radio Discussion List ListServ, March 4, 1996)

As discussed in Chapter 3, in regard to forfeitures (the station's being fined), the notion that the college radio station is staffed by nonprofessional, untrained (and in more recent years, unlicensed) novices (known as students) is almost frightening. In reality, noncompliance with mandated regulations is not justified due to a lack of training or ignorance. Here, the idea that the radio station is a training facility is not a consideration. Adherence to rules and regulations is expected by noncommercial educational stations, just as it is with their commercial counterparts.

Also mentioned in Chapter 3 was the station public file. "As part of its obligations as a public trustee, a noncommercial broadcast licensee must make certain materials and records available for public inspection and copying. Failure to maintain or update a public file can result in FCC penalties" (National Association of College Broadcasters 1995, 76). This file must be available to the general public during normal business hours.

> The following materials and documents must be placed in the Public File and kept for the time period indicated. You should review Section 73.3527 for a more complete discussion of each of these items.
>
> 1. FCC Applications.
> 2. Ownership Reports.
> 3. Contracts and Agreements.
> 4. Requests for Political Broadcast Time.
> 5. Annual Employment Reports.
> 6. Issues-Programs Lists.
> 7. FCC Procedure Manual.
> 8. Donor List.
> 9. Station-FCC Correspondence.
>
> (National Association of College Broadcasters 1995, 98)

(There is an ongoing question within noncommercial educational stations if the Issues/Programs Lists are required for inclusion in the public file. Much of the confusion concerns "exempted" Class D stations.)

Here, again, the station manager/advisor and chief engineer need to keep abreast of requirements. As of 1998 the FCC was moving toward updating public file requirements to include such items as contour maps and complete copies of all license authorizations. Additionally, the FCC Procedures Manual was being updated. As of June 1999 "The Public and Broadcasting," which is required to be kept in the public file, was available from the FCC Web site at: *www.fcc.gov/mmb/prd/docs/manual.wp*. Futuristically, posting of the public file on the World Wide Web was being encouraged.

> *[Our station] will undergo its first license renewal application this academic year. We are up for renewal in Aug. 98. Can anyone tell me what I must do to prepare for this? Do I need to contact our lawyer? (National Association of College Broadcasters-Faculty/Staff ListServ, August 28, 1997)*

As is to be expected, the sole responsibility of the operation of the college radio station rests with the licensee—the school itself. But adherence to rules and regulations falls under the guidelines and oversight of the station manager (again, usually either a faculty member or staff employee), who has been delegated to "run the station." (See Jim McCluskey's 1998 article detailing new and existing FCC rules and policies that affect college radio stations. This provides a good example of the importance of station managers and engineers keeping up-to-date on regulatory issues.)

Included is the responsibility of ensuring that the station operates with the standards of good engineering practice as set forth in part 73.508 of the FCC Rules and Regulations (Rules Service Company, 1994-1995, Record 2413-2416/662; Fig. 6.1). Current self-inspection checklists from the FCC can be accessed via its Web site at: *http://www.fcc.gov/cib/Publications/fm.html*. (Use *am.html* for AM Standard Broadcast stations or *tv.html* for television stations). Additionally, one can access the FCC Compliance and Information Bureau at *http://www.fcc.gov/cib/bc-chklsts/Welcome.html*, which will also lead the reader to self-inspection checklists. Finally, stations can inquire with state broadcast associations as to alternative inspection programs offered.

Also, if the station or school owns its tower, an additional regulatory agency comes into play: the Federal Aviation Administration (FAA). Again, this is where keeping up with the rules is a necessity (i.e., tower registration and lighting requirements).

Sources for legal guidance include the *NAB Legal Guide* (available from the National Association of Broadcasters). The NAB offers a discounted membership rate for college stations. The NAB home page is *www.nab.org*.

RADIO STATION (Call Sign)

Regulations & Restrictions

These Programming Regulations and Restrictions have been adopted to ensure that Radio Station ____ broadcasts programs of the highest possible standard of excellence and in compliance with the Rules and Regulations of the Federal Communications Commission.

I. CONTROVERSIAL ISSUES. Any discussion of controversial issues of public importance shall be reasonably balanced with the presentation of contrasting viewpoints in the course of overall programming; no attacks on the honesty, integrity, or like personal qualities of any person or group or persons shall be made during the discussion of controversial issues or public importance; and during the course of political campaigns, programs are not to be used as a forum for editorializing about individual candidates. If such events occur, the Licensee may require that responsive programming be aired.

II. NO PLUGOLA OR PAYOLA. The following business activities or "plugs", relating to the payment, acceptance of payment, agreement to pay or agreement to accept payment of money or other consideration is prohibited: (a) taking money, gifts or other compensation from any person for the purpose of playing any record or records on the air; (b) taking money, gifts or other compensation from any person for the purpose of refraining from playing any record or records on the air; (c) taking money, gifts or other compensation from any person for the purpose of promoting any business, charity or other venture without first informing the Station's General Manager, and (d) promoting any business venture which is unconnected with Radio Station _____ on the air without first informing the General Manager.

III. ELECTION PROCEDURES. At least ninety (90) days before the start of any primary or regular election campaign, the Station's General Manager will determine the rate to charge for underwriting time to be sold to candidates for the public office and/or their supporters to make certain that the rate charged is in conformance with the applicable law and station policy.

IV. PROGRAMMING PROHIBITIONS. The Station shall not broadcast any of the following programs or announcements:

A.   **False Claims.** False or unwarranted claims for any product or service.

B.   **Unfair Imitation.** Infringements of another advertiser's rights through plagiarism or unfair imitation of either program idea or coy, or any other unfair competition.

C.   **Commercial Disparagement.** Any disparagement of competitors or competitive goods.

D.   **Indecency.** Any programs or announcements that are slanderous, obscene, profane, vulgar, repulsive or offensive, either in theme or in treatment.

E.   **Unauthenticated Testimonials.** Any testimonials which cannot be authenticated.

F.   **Descriptions of Bodily Functions.** Any continuity which describes, in a patently offensive manner, internal bodily functions or symptomatic results of internal disturbances, or reference to matters which are not considered acceptable topics in social groups.

V.   **No Lotteries.** Announcements giving any information about lotteries or games prohibited by federal or state law or regulation are prohibited. This prohibition includes announcements with respect to bingo parties and the like which are to be held by a local church, unless expressly permitted by State law.

VI.   **No "DREAM BOOKS."** References to "dream books," the "straight line", or other direct or indirect descriptions or solicitations relative to the illegal numbers lottery, "numbers game," or the "policy game," or any other form of gambling are prohibited.

FIGURE 6.1 Radio Station Regulations and Restrictions
Source: Booth, Freret, Imlay & Tepper, P.C. Used with permission.

VII.   **No NUMBERS GAMES.** References to chapter and verse numbers, paragraph numbers, or song numbers which involve three digits should be avoided and, when used, must be related to the overall theme of the program.

VIII.   **NO CASINO GAMBLING.** The broadcast of information which promotes the patronizing of gambling casinos is prohibited.

IX.   **NO INDIAN BANKING CARD GAMES.** The mention of Indian banking card games such as baccarat, chemin de fer, and blackjack, as well as electronic or electro-mechanical facsimiles of any game of chance or slot machines, casino gambling, craps, roulette and betting parlors is prohibited.

X.   **NO OFF-RESERVATION INDIAN RAFFLES.** The mention of any Indian gaming, such as rate, conducted off Indian lands, is prohibited.

XI.   **RELIGIOUS PROGRAMMING RESTRICTIONS.** Any religious programming broadcast is subject to the following restrictions:

A.   **Respectful of Faiths.** The subject of religion and references to particular faiths, tenets, and customers shall be treated with respect at all times.

B.   **No Denominational Attacks.** Programs shall not be used as a medium for attack on any faith, denomination, or sect or upon any individual or organization.

C.   **Donation Solicitation.** Requests for donations in the form of a specific amount (for example, $1.00 or $5.00), shall not be made if there is any suggestion that such donation will result in miracles, cures or prosperity. However, statements generally requesting donations to support the broadcast or the church are permitted.

D.   **Treatment of parapsychology.** The advertising or promotion of fortunetelling, occultism, astrology, phrenology, palm reading, or numerology, mind-reading character readings, or subjects of the like nature is not permitted.

E.   **No Ministerial Solicitations.** No invitations by the minister or other individual appearing on the program to have listeners come and visit him or her for consultation or the like shall be made if such invitation implies that the listeners will receive consideration, monetary gain, or cures for illness.

F.   **No Miracle Solicitation.** Any invitations to listeners to meet at places other than the church and/or to attend other than regular services of the church is prohibited if the invitation, meeting, or service contains any claim that miracles, cures, or prosperity will result.

X.   **CREDIT TERMS ADVERTISING.** Pursuant to rules of the Federal Trade Commission, no advertising of credit terms shall be made over the Station beyond mention of the fact that if desired, credit terms are available.

XI.   **NO ILLEGAL ANNOUNCEMENTS.** No announcements or promotion prohibited by federal or state law, or regulation of any lottery or game, shall be made over the Station. Any game, contest, or promotion relating to or to be presented over the Station must be fully stated and explained in advance to the Licensee, which reserves the right in its sole discretion to reject any game, contest, promotion.

XII.   **LICENSEE DISCRETION PARAMOUNT.** In accordance with the Licensee's responsibility under the Communications Act of 1934, as amended, and the Rules and Regulations of the Federal Commissions, the Licensee reserves the right to reject or terminate any programming or sponsorship announcement proposed to be presented or being presented over the station which is in conflict with Station policy or which in the Licensee or its General Manager/Chief Engineer's sole judgment would not serve the public interest.

XIII.   **FOREIGN LANGUAGE PROGRAMS.** The foreign language broadcast of any programs, commercials, announcements, PSA's or other content, where the English translation thereof would violate any restriction contained herein, is prohibited.

FIGURE 6.1 (*continued*)

Additionally, as noted within Chapter 2, the most recent addition of the Code of Federal Regulations can be accessed online at: *htt://www.access.gpo.gov/nara/cfr/index.html*. The Audio Services Division of the FCC can be accessed via *http://www.fcc.gov/mmb/asd/*.

> Although university broadcast stations are exempt from some of the regulations facing their commercial counterparts, a number of legal concerns must still be addressed by the faculty advisor. Apart from complying with FCC regulations, faculty must be aware of general communication laws so that the university's broadcast license may be protected. These laws are also important components in the students' education since, upon graduation, they will be expected to navigate in the "real world" of broadcasting. It is essential that the faculty advisor have a basic understanding of the basic legal issues facing university stations and where to get answers to questions that will inevitably arise. (Creech 1996, 5)

Station Policy Manuals

From the outset, station policy manuals serve two purposes: (1) they force station management to outline the intent and operation of the station; and (2) they help to provide needed continuity. Remember (it's worth repeating!): Turnover at the campus radio station is guaranteed!

> The advantages of a staff manual clearly outweigh the obstacles to making it happen. By committing your policies to paper, you have a definitive text of all station operation procedures and policies. You can ensure a swift, orderly, and complete dissemination of knowledge critical for your new staff to have about how your station "works." This is especially crucial today: after the deregulation of the '80s, all operators and licensees now have increased responsibility to comply with the remaining FCC rules and regulations. (National Association of College Broadcasters 1995, 102)

Usually included in the station policy manual is the mission and/or goals. It might not be a bad idea to include an established station philosophy, which can be adapted and used in other areas, such as your station Web page (an example is included Chapter 3). A historical perspective of the station is useful, including names of those who worked at the station as students and are now well-known professionals in the broadcasting industry. Other items to consider are a station coverage map and any listener facts and figures available concerning the station audience (or intended audience).

> *I think the mission of noncommercial educational radio varies greatly from station to station. It also depends on how the station is funded. For the most*

part, college stations that are funded by student government funds and the like seem to be more free-form, play-whatever-you-want stations. Stations like the one I run, which are funded by a communications department or something similar are treated more like laboratory stations and used to prepare students for commercial radio careers. (National Association of College Broadcasters ListServ, September 29, 1997)

Of course, the basic organizational structure (even an organizational chart, if possible) of the station should be outlined in the station policy manual. This allows for the new staff member to determine who to go to for help. If the manual is truly up-to-date, names can actually accompany staff-director positions.

All station-operating policies need to be laid out in the station policy manual. Besides basic FCC rules and regulations (obscene/indecent programming, payola/plugola, underwriting policies, broadcasting telephone conversations, etc.) and school policies (no smoking in buildings, no alcohol on campus, volunteer agreements [sample provided in Fig. 6.2], etc.), specifics down to how to answer the telephone and when to call the chief engineer at home should be included. "While these topics are indeed the minutiae of daily organizational life, they seem to fit the '80-20 rule': 80% of your headaches will come from this 20% of your policy book." (National Association of College Broadcasters 1995, 103)

Let the station policy manual be a true working document. Although it will take a great deal of time to first put it together, in actuality most of it has prob-

KNTU-FM VOLUNTEER AGREEMENT

This agreement covers volunteer labor donated to KNTU-FM, Denton, Texas by _____
                                                                                    (your name)

in the position of _____ commencing _____.
                    (job you are volunteering for)   (today's date)

I agree to work in a non-paying, volunteer position at KNTU-FM. I understand that this is a learning experience to assist my career development and that no additional compensation will be given. I further understand that I will be subject to the rules, regulations, and policies of the University and KNTU-FM (see KNTU Policy Manual in newsroom or on-air studio), as are regular student employees, and that I will be evaluated in accordance with those rules and regulations.

This agreement will remain on file at KNTU-FM. This agreement covers only volunteer work for the position named, and shall remain in effect as long as this student continues to hold the position named herein. Any other position held by the student/participant at KNTU-FM will be covered under separate agreement. Sign below:

_____   _____
STUDENT/PARTICIPANT SIGNATURE   DATE

FIGURE 6.2 University of North Texas KNTU-FM Volunteer Agreement Form
Source: KNTU-FM, University of North Texas. Used with permission.

ably been written before in memos and station meeting minutes. It just needs to come together in a coherent format. And once it is put together, keep it up-to-date. This is where management will really determine just how specific it wants to get. The more precise the information, the more that information will need to be reviewed and updated.

The station policy manual can also be the place where the station can articulate its position on such philosophical ideals as editorializing within station broadcasts. Some stations have been known to adopt guidelines presented in such publications as the *Statement of Principles of Editorial Integrity in Public Broadcasting* (Editorial Integrity Project, 1986) and blend those into their station manual.

If it's possible, have someone outside of the station, but with broadcasting knowledge, review the manual. These are the people who can call attention to areas overlooked in the manual.

Then, when the station policy manual is ready, either provide copies of the manual to your staff, or at least have it located in a place that is easily accessible, in a general area. The manual should *not* sit on the manager's office shelf.

Station Staff Responsibilities

I do know for a fact that college radio is an asset if you want a career in broadcast journalism. That doesn't mean it's not for a disc-jockey career. But being a disc-jockey is a different skill that requires the ability to entertain and be a personality. It also requires practice in a real station environment. Some colleges can provide that and do. Graduates from those schools are always going to do well compared to the ones that run a station like a hobby. (Airwaves Radio Journal ListServ, Issue #1582, October 11, 1995)

Whereas the policy manual details the makeup of the station staff, in this section actual responsibilities are detailed. This, then, lays out who runs the college radio station from the inside. It should also help to point out that there are many more individuals working at the station off the air than are actually broadcasting. These are only general descriptions of the positions.

Professional/Academic Full-Time

● *General Manager or Station Manager.* This individual is quite often the faculty advisor to the station, with station duties assigned in addition to academic (teaching) responsibilities. At some schools this person is a full-time staff member. It is the responsibility of the person to oversee the management and administration of the entire radio station. The manager has complete control over the station and maintains total and legal responsibility for its operation. All personnel, budgetary, and general programming decisions are under the auspices of the general manager/station manager.

• *Chief Engineer.*   Usually this individual is a full-time staff member, possibly dividing his or her time with other duties within an academic department or unit. Recent rulings by the FCC may now have this position labeled as the chief operator, as delegated by the station manager. This person is responsible for all technical specifications and operations at the station, including studio and broadcast.

Student Positions

(Some college radio stations, particularly NPR and commercial stations, fill a few of these positions with full-time personnel.)

• *Student Station Manager.*   Oversees the day-to-day operation of the station. Supervises the entire student staff.

• *Program/Operations Director.*   In lieu of the student station manager, the highest-ranking student position. Oversees the entire day-to-day operation, with more concentration on station programming; auditions and schedules all disc jockeys.

• *News Director.*   Supervises all news operations within the station, including the auditioning and scheduling of newscasters and reporters. May oversee public affairs programming.

• *Sports Director.*   Supervises all sports operations within the station, including the auditioning and scheduling of sportscasters and sports reporters. Also coordinates all remote sports broadcasts.

• *Public Affairs Director.*   Works in conjunction with the news director to assume responsibility for all public affairs production and programming. May also assist in the ascertainment of community leaders to determine programming in accordance with the quarterly issues/programs lists.

• *Music Director.*   Assumes responsibility for all music played at the station. Maintains contact with music representatives. Coordinates giveaways, promotions, and live broadcasts with the program director and promotions director.

• *Traffic Director.*   Works closely with the program director while maintaining the station program logs. Ensures that requirements are met concerning EAS, underwriting, and other legal obligations. May be responsible for ensuring that programs for broadcast are in the on-air studio in a timely fashion.

• *Production Director.*   Assumes responsibility for all station production, in particular spot (short form :30, :60) production. Works closely with the program, news, sports, and public affairs directors to ensure production quality.

- *Public Service Director.* Ensures that public service obligations are met through the airing of public service announcements. Works closely with the program and promotions directors in coordinating station public service activities, particularly live remote events. In conjunction with the production director, prescribes the recorded/produced public service programming on the station.

- *Promotions Director.* Handles all station promotions. Works closely with the program and music directors to coordinate both in-studio and remote station promotions. Works with the station manager, program director, and sales director to produce the station program guide for outside distribution.

- *Sales Director.* Seeks out, solicits, and services underwriting/supporting clientele. Works closely with the station manager and program director to ensure that legal and programming standards are met.

- *Special Programs Director(s).* Takes on tasks usually specific to a given formatting area (i.e., alternative, country, new-age, urban, and so on). Quite often coordinates programming within that area, including the auditioning and scheduling of special programs hosts.

Perspectives of a College Radio Workforce

This section includes references to works that examine the area of station administration and staffing. Again, these sources are meant to serve as a starting point for the reader's own research.

The reader is first directed to a dissertation written in 1993 by Charles G. Bailey entitled *Perceptions of Professional Radio Station Managers of the Training and Experience of Potential Employees Who Have Worked in College Radio under One of Three Different Administrative Patterns.* Of particular interest here is the section "Educational Radio Stations Owned and Operated by Colleges and Universities" (100-119), which provides a discussion of management and administration at stations. I also recommend that the summary (262-67) and the summary of conclusions (285-96; 200-201) be reviewed. Bailey writes:

> The likelihood of the thirteen skill areas being acquired by students under the supervision of either a faculty manager, a staff manager, or a student manager was determined by the commercial radio station managers' ratings of likelihood and the resulting mean scores of the composite ratings. The top five skill areas perceived as likely to be acquired by students under the supervision of a faculty manager are writing, communication, legal and regulatory aspects, studio equipment operation, and daily operational tasks. The top five skill areas perceived as likely to be possessed by students under a staff manager are studio equipment operation, daily operational tasks, communication, writing, and legal

and regulatory aspects. The top five skill areas perceived as likely to be acquired by students under a student manager are studio equipment operation, daily operational-tasks, remote equipment operation, computer, and telephone. The thirteen skill areas and the likelihood mean scores for each of the three administrative patterns may be found in Table [6.1]. (286-87)

Also of note is Dennison's 1994 study *Administrative Patterns of On-Campus Radio Stations and the Leadership Behaviors of the Managers.* Dennison observes that:

> A station manager at a public college or university will have to deal with a number of different offices and agencies in order to execute day to day station operations. A manager might have to clear up a problem with a travel budget through a state approved travel agency, deal with approved bid specifications from the purchasing department, coordinate fund raising through the office of

TABLE 6.1.   **Summary of the Commercial General Managers, Perceived Importance of Skills for Entry-level Positions and the Likelihood of Students Acquiring These Skills Having Worked in College Radio Under the Supervision of a Faculty Manager, Staff Manager, or Student Manager, in Order of Importance**

Entry-level skill area	Importance[a] of the skill area	Likelihood[b] of skill area under the supervision of a		
		Faculty manager	Staff manager	Student manager
Communication	1.10	1.95	2.07	3.11
Telephone	1.30	2.54	2.63	2.96
Writing	1.36	1.91	2.10	2.98
Studio equipment operation	1.45	2.08	1.75	2.33
Daily operational tasks	1.50	2.13	1.91	2.40
Legal and regulatory aspects	1.75	1.95	2.12	3.19
Marketing techniques	1.79	2.52	2.41	3.54
Management	1.87	2.19	2.14	3.00
Remote equipment operation	1.88	2.49	2.18	2.71
Computer	1.96	2.21	2.40	2.72
Typing	2.13	2.54	2.63	2.96
Engineering	2.68	3.01	2.88	3.56
Minimum foreign language proficiency	3.88	3.14	3.50	3.79

Note: Mean scores are rounded off to two decimal places.
[a]Importance means values indicate: 1 = Very important, 2 = Somewhat important, 3 = No opinion, 4 = Somewhat unimportant, 5 = Very unimportant.
[b]Likelihood means values indicate: 1 = Very likely,   2 = Somewhat likely, 3 = No opinion, 4 = Somewhat unlikely, 5 = Very unlikely.

Source: Bailey, 1993, 200-201. Used with permission.

institutional affairs and seek additional operating money from the vice president for business affairs.

A station manager at a private college or university is closer to the source of funding and therefore operates in a different and perhaps less formalized system of communication. (59)

Internship Opportunities

Programming Department
- Assist in Production
- Assist in Music Research
- Audience Research
- Assist in Programming Special Features
- Prize Coordinator
- Board Operation Procedures
- Music Scheduling (playlist)
- Assist Liner Coordinator

Marketing/Promotions Department
- Graphics Design
- Video Production
- Public Relations
- Copy Writing
- Convention Planning
- Administrative Duties Involved
- Business World Realities
- Marketing Concept Campaign (direct mail)
- Marketing Strategies
- Advertising Copy/Headline Writing
- Photo Ops
- Print Production
- Desk Top Publishing

CONTACT: Kathryn Green
972/448-3305

13725 Montfort Drive
Dallas, TX 75240

abc ABC RADIO NETWORKS

FIGURE 6.3 ABC Radio Networks Internship Opportunities
Source: ABC Radio Networks. Used with permission.

Of course, all of the proper training and operation of the college radio station leads to producing "professional broadcasters" out of student DJs. This, then, leads to losing your staff to internships (see Fig. 6.3) and real jobs. But it must be always remembered that a true purpose of college radio is to help provide tomorrow's broadcasters. Thus, how and who runs the station plays an important role in this endeavor.

REFERENCES

Bailey, C. G. (1993). Perceptions of professional radio station managers of the training and experience of potential employees who have worked in college radio under one of three different administrative patterns. Ph.D. diss., West Virginia University, 1993. Abstract in *Dissertation Abstracts International,* A54/08:2809.

Creech, K. C. (1996). Up and running: Legal considerations in operating university broadcast facilities. Paper presented at the 1996 Broadcast Education Association Meeting, Student Media Advisors Session, April 13, 1996, Las Vegas, Nev.

Dennison, C. F., III. (1994). Administrative patterns of on-campus radio stations and the leadership behaviors of the managers. Ph.D. dissertation, West Virginia University, 1994. Abstract in *Dissertation Abstracts International* 54:3940A.

Editorial Integrity Project. (1986). *Statement of principles of editorial integrity in public broadcasting.* Columbia, S.C.: Southern Educational Communications Association.

Hamilton, C. E. (1994). The interaction between selected public radio stations and their communities: A study of station missions, audiences, programming and funding. Ph.D. diss., University of California, San Diego, 1994. Abstract in *Dissertation Abstracts International* 54:3025A.

Holterman, S. (1992). The relationship between record companies and college music directors: A descriptive study of alternative radio. Master's thesis, University of Tennessee, Knoxville.

Kruse, H. C. (1995). Marginal formations and the production of culture: The case of college music. Ph.D. diss., University of Illinois, Urbana-Champaign, 1995. Abstract in *Dissertation Abstracts International,* A56/09:3360.

Leidman, M. B. (1985). At the crossroads: A descriptive study of noncommercial FM radio stations affiliated with colleges and universities of the early 1980s. Ph.D. diss., George Peabody College for Teachers of Vanderbilt University, 1985. Abstract in *Dissertation Abstracts International,* A46/06:1604A.

McCluskey, J. (1998). An examination of new and existing FCC rules, policies and procedures affecting student/noncommercial radio stations. *Feedback* 39(4): 32-36.

McDonald, G. (1995). Left of the dial. *U. Magazine,* April, 20-21.

National Association of College Broadcasters. (1995). *1995 NACB station handbook.* Providence, R.I.: National Association of College Broadcasters.

Ozier, L. W. (1978). University broadcast licensees: Rx for progress. *Public Telecommunications Review* 6(5): 33-39.

Poole, C. E. (1989). Market positioning theory: Applicability to the administration of public radio stations operated by institutions of higher education. Ph.D. diss., University of Florida, 1989. Abstract in *Dissertation Abstracts International* A51/03:757A.

Reese, D. E. (1996). College radio from the view of the student staff and the audience: A comparison of perceptions. *Feedback* 37(2): 17-19.

Rules Service Company. (1994-1995). *Part 73 radio broadcast services.* Rockville, Md.: Rules Service Company.

Ryan, L., and S. Baruth. (1993). From the classroom to the boardroom: A Gavin special focus on the college radio route into the music business. *Gavin,* May 14, 40-43.

Sauls, S. J. (1993). An analysis of selected factors which influence the funding of college and university radio stations as perceived by station directors. Ph.D. dissertation, University of North Texas, 1993. Abstract in *Dissertation Abstracts International* 54:4372.

————. (1995). College radio. Entry submitted to the *Encyclopedia of United States Popular Culture.* Santa Barbara, Calif.: ABC-Clio, forthcoming.

————. (1996). College radio: Points of contention and harmony from the management perspective. *Feedback* 37(2): 20-22.

————. (1997). Who's running college radio? Paper presented within the Radio Interest Group at the 1997 Popular Culture Association/American Culture Association National Conference, San Antonio, Texas, March 28, 1997. ERIC, ED 411 561.

Thompsen, P. A. (1992). Enhancing the electronic sandbox: A plan for improving the educational value of student-operated radio stations. *Feedback* 33(1): cover, 12-15.

Wilkinson, J. S. (1994). College radio: Farm team or free form funhouse? *Feedback* 35(1): 4-7.

Wood, D. N., and D. G. Wylie. (1977). *Educational Telecommunications.* Belmont, Calif.: Wadsworth.

Who's Paying
for College Radio?

T hat college stations must please the entities that fund them is a harsh reality of the business. As noted in Chapter 5, "although some college stations have switched to a top 40 format and emulate professional stations, most are still eclectic, noncommercial, and proud of it. But to stay afloat, and to grow, they must please their sponsoring campus groups" (Knopper 1994, 84). Or be subject to possible budget cuts, or even suspension. But, as Stephen Fisher, new music program coordinator at the University of San Francisco's KUSF, says:

> Providing "cultural programming," winning awards, avoiding radio violations, and operating with a sense of "what they don't know won't hurt them" staves off the budget cutters.
>   Basically, they leave us alone, Fisher says. More because they don't understand us, not because they want to leave us alone. (Knopper 1994, 84)

This idea is ever present in the funding of campus radio stations. The question of loyalty, campus politics, playing favorites, and servicing numerous groups all come into play, helping to secure, increase, or diminish station support. And, as mentioned in Chapter 6, control and operation of the station can be directly tied to the station's funding source. This chapter, therefore, deals specifically with the funding of college and university radio stations.

My own research in my doctoral studies concerned this subject. My dissertation, entitled *An Analysis of Selected Factors Which Influence the Funding of College and University Noncommercial Radio Stations as Perceived by Station Directors* (Sauls 1993), focused entirely on station funding. A great deal of the material in this chapter grew from that original research (see Sauls 1998).

FUNDING THE COLLEGE RADIO STATION

From the outset it can be stated that, generally, the funding of college radio is limited. Funding sources for college and university radio stations vary greatly, with the bulk traditionally coming from student fee support or general academic funds. In his book entitled *The College Radio Handbook,* Brant (1981) points out that "with few exceptions college radio stations are budgeted by the college or university to which they are licensed" (82). Brant also notes that the few commercial stations licensed to educational institutions have been able to seek potential advertisers (84). Some of these stations, such as Howard University's WHUR-FM, have even achieved superiority ratings in major markets (Evans 1986).

In 1979 Lucoff lamented that, while university administrators generally have little or no broadcasting experience, they most often possess "control over funding" of campus radio stations (26). In contrast, however, the National Association of Educational Broadcasters reported that 75 percent of the college and university educational radio stations responding to their 1967 study had only monthly or less frequent contact with the college or university as the licensee in the operation of the station (I-14). It is "hardly surprising to find a direct connection between budget size and the quality and extent of station programming" (I-8). Thus, it might be wise for those managing and advising such stations to communicate directly with school administrators who oversee funding. A little in-house "bragging" in regards to station potential on campus can go a long way! (see Sauls 1996, 20).

The limitation factor of funding forces many stations to exert a lot of energy to raise external support. These efforts might include soliciting outright cash or equipment donations, on-air "auctions" of donated items, on-air pledge drives, T-shirt sales, and even sales of station promo compact discs (although the legality of this may be questionable). The point is, the station must measure the amount of work involved and determine its best option to raise money. Various other options are discussed in this chapter, including noncommercial underwriting and grant support.

CPB AND STATION FUNDING

> *First of all, it could be argued (but not by me) that Newt [Gingrich] was the best thing to ever happen to public radio. Faced with the threat that CPB funding would be "zeroed out" under the new Congress, stations that pushed hard for new members and larger gifts generally got both. Local income and membership soared. (Personal communication, November 8, 1998)*

As discussed in Chapter 3, programming entities such as National Public Radio can supplement a great majority of a station's programming. Additionally, these groups can provide much-needed funding. The question, of course, arises regarding the funding source, programming control, and loyalty.

In order to better understand such funding, this chapter provides an explanation of the structure under which National Public Radio operates. In particular, the umbrella aspect of the Corporation for Public Broadcasting is outlined. Created by the Public Broadcasting Act of 1967, the Corporation for Public Broadcasting, or CPB, is a "private, nonprofit corporation that oversees the distribution of the annual Federal contribution to the national public broadcasting system" (Corporation for Public Broadcasting 1995-96). In addition to funding diverse and innovative radio and television programs (educational, informational, and cultural), CPB distributes grants to CPB-supported public radio stations (National Public Radio, which began in 1970) and to public TV stations (Public Broadcasting Service, created in 1969). Sources of primary funding for the Corporation for Public Broadcasting are congressional appropriations, foundations, and corporations (Corporation for Public Broadcasting 1995-96).

A 1996 report entitled *Public Broadcasting's Service for the American People,* from the Corporation for Public Broadcasting, reviewed the services that had been undertaken by CPB and public broadcasters: "They include important contributions in the areas of education, community service, and technology. These services may be less well known than the radio and television programming that helps keep millions of Americans informed, but they are no less important a part of public broadcasting's mission" (Corporation for Public Broadcasting 1996, 3).

Specifically, concerning funding, the report states that "CPB grants are especially important to stations serving rural audiences. For these stations, which may have a smaller base of outside support, CPB grants can represent almost one-third of the operating budget" (7). It continues that "CPB funds represent 16% of public radio's overall income, providing the foundation for other fundraising activities. The median Community Service Grant to an individual public radio station is just a bit less than $100,000" (8).

What about the disappearing federal dollars? Well, they did diminish for some stations. But because the federal grants are determined by the amount of money raised by each individual station, those stations that experienced huge increases in listener donations were often able to keep the same size grants—or even increase them. That is not to say that stations did not feel a pinch. They did, but mostly because of concurrent huge increases in program costs. (Personal communication, November 8, 1998)

Of note, in 1995 the federal funding of the Corporation for Public Broadcasting was under congressional scrutiny. A good number of college radio and television stations are NPR or PBS affiliates, which rely on CPB funds (Petrozzello 1995, 50). It is possible, therefore, that a reduction of such federal funding could impact campus media programming and support. To illustrate this fact, as of March 29, 1996, it was reported that actual fiscal 1995 Corporation for Public Broadcasting funding was $286,000,000, fiscal 1996 estimate was $275,000,000, and the fiscal 1997 request at that time was slated at $260,000,000 (Clinton's fiscal 1997 budget plan, 1996, A43). This shows a decline in CPB current and anticipated funding brought on by the fact that, as some believe, the "politicians denied the system a stable funding source, forced it to increasingly depend on corporate largess, and assaulted its presumption to intelligence and independence from the status quo" (Landay 1996, 19). In sum, the picture portrays "angry congressional budget-cutters deriding the need for public broadcasting and complaining about liberal bias on the air; states cutting back on their contributions; [and] university licensees following suit while sometimes demanding greater on- and off-air identification with their station" (Conciatore 1995, 17).

> *Though we did increase our income from listeners, I truly believe damage was done. For the first time, we raised the theory that perhaps federal funding was not a good idea for public radio. Many managers and listeners began to believe that we had better become more independent. We began to dispose of the notion that there is a "public responsibility" to support the mission of public radio. Many managers today sound much like commercial station managers. The experience has hardened many of us. It has made us think of money in ways formerly appropriate only for commercial managers. We ask ourselves questions about how we can reach more listeners—so that we can get more donations—or more underwriting. We have scuttled much of our "unpopular" programming, which is perceived as unable to support itself in the "market-place." (Personal communication, November 8, 1998)*

UNDERWRITING AND COMMERCIAL POLICIES

The problems associated with declining government funding were again noted in 1998: "Declines in federal and other public funding for non-coms [noncommercial stations] during the last decade have forced public radio to invest resources in finding and keeping both individual contributors and business sponsors or underwriters" (Yana Davis 1998, 17). This, then, leads to the discussion of program underwriting/sponsorship and commercial advertising at the college radio station.

The National Association of College Broadcasters' *1995 College Radio Survey* provides information that could be of importance to current and potential sponsors of college radio programming. This study specifically addresses program support via underwriting and outright commercial advertising. (For particulars regarding underwriter announcements see the practical applications later in this chapter.)

The study shows that, generally, stations have far fewer restrictions on running underwriting announcements than they do commercial spots. Specifically, 61 percent of the respondents reported that their station has no restrictions in its use of underwriting (see Table 7.1), compared to 68 percent who are restricted by their license from running commercial spots (see Table 7.2). Twelve percent report that they have no restrictions from running commercial spots. This compares to 3.9 percent of the stations restricted by institutional policy from running underwriting spots.

It should be noted that the NACB (and others) have tried to initiate national underwriting and commercial advertising campaigns on college stations. Audience size, listenership estimates, and the concern governing underwriting versus commercials on college radio all have played a role in such endeavors. Past efforts included the NACB's Interep Cooperative and the National Student Radio Co-operative. Additionally, contemporary business services may be used to help solicit underwriting. For example, the College Radio Advertising Network assists noncommercial college radio stations in seeking out corporate sponsors who are searching for college audiences. Their Web site is located at *http://www.cranetwork.com.*

Table 7.1   Range of Underwriting Policies (*n*=228)

Underwriting Policy	*Percentage Responding*
Restricted by institutional policy from running any underwriting	3.9%
Limited by institutional policy to running some underwriting	9.2
Restricted by internal/organizational policy from running any underwriting	.9
Limited by internal/organizational policy to running some underwriting	9.6
No restriction in our use of underwriting	60.5

Note: Since some respondents did not answer any of the underwriting questions, responses do not add to 100 percent.

Source: National Association of College Broadcasters. (1995a) *1995 College Radio Survey*, P. 32. Used with permission.

Table 7.2   Range of Commercial Policies (*n* = 228)

Commercial Policy	*Percentage Responding*
Restricted by license from running any commercial spots	68.4%
Restricted by institutional policy from running any commercial spots	14.5
Limited by institutional policy to running some commercial spots	2.6
Restricted by internal/organizational policy from running any commercial spots	9.2
Limited by internal/organizational policy to running some commercial spots	3.5
No restriction in our running of commercial spots	12.7

Note. Since respondents could check more than one response, percentages add to more than 100 percent.

Source: National Association of College Broadcasters. (1995a) *1995 College Radio Survey*, P. 31. Used with permission.

Finally, a point should be made that, at the time of this writing, low-power stations that fall under part 15 of the FCC rules may be entitled to air commercial advertisements. As always, any programming decision of this nature should first cleared by legal counsel.

And these days we often hear the program underwriting "business" much more openly referred to using commercial terms—like "avails" and "spots" and "rate cards." Some of those underwriting announcements sound dangerously close to advertisements. Some run 30 seconds or more. My fear is it could all collapse inward if our audience begins to feel we are no different from our commercial brethren, and that we no longer need listener support. (Personal communication, November 8, 1998)

PRACTICAL APPLICATIONS

Membership Solicitation

Stations also solicit outright cash donations for support. Figure 7.1 provides an example of a membership renewal solicitation notice. Features such as contests and sweepstakes, automatic renewals, special discounts (local retailers, attractions, restaurants, etc.) for members, and direct bank account plans can all enhance membership pledge efforts.

Concerning member subscriptions, "[f]aced with cuts in government funding, noncommercial [NPR supported] stations are re-examining their music mix. They must determine how to draw in those listeners who will ultimately subscribe to the station, while at the same time, try to avoid alienating current fans" (Gronau 1995, 22). In sum, programmers do have to take into account listener support when determining format trends and changes.

Grants

Our campus station, _____-FM, is looking for any information related to grants that might be available to support the production of programming materials and the syndication of programming materials. If anyone knows of any grants available in these areas, I'd appreciate the information. (Personal communication, NACB ListServ, March 27, 1998)

From the outset, let me say that securing a grant can be a very long process. The recipient of a Public Telecommunication Facilities Program grant (NTIA/PTFP can be located on the Web at *www.ntia.doc.gov*) for my university radio station in 1986, I experienced approximately a one-year period of grant preparation and two one-year cycles with PTFP to obtain the grant. But it was well worth the effort. The following discussion is presented to foster the ideals of securing grants for station support.

Dear Member and Friend,

    Since we announced the change in KERA 90.1 membership, I have been very encouraged by the enthusiastic response.

    Our members understand that making all memberships coincide with the regular calendar year -- rather than renewing members at different times all throughout the year -- makes a lot of sense.

    It makes <u>practical sense</u>, because we can handle all our renewals at once.

    And it makes <u>financial sense</u>, because we'll save money on renewal mailings, processing and so on.

    And you seem to like it!

    Many KERA 90.1 members have already renewed their memberships for 1999.

    If you are one of them, thank you. But, if not, won't you take this opportunity to send in your renewal now?

    You know, a well-conceived change in how we run the station can be a very good thing. In this case, we'll be better able to plan our budgets, and we'll save money that we can then spend on the programs you want.

    But, one thing that will <u>never change</u> here at KERA 90.1 is our determination to program the station to <u>your needs and taste</u> for complete, in-depth news ... intelligent, informative conversation ... and entertainment you just won't hear anywhere else.

    Where else but on KERA 90.1 will you find programs like <u>Morning Edition</u> ... <u>All Things Considered</u> ... <u>Fresh Air</u> ... and <u>Car Talk</u>?

    Personalities like Sam Baker, Glenn Mitchell, Diane Rehm and Ray and Tom Magliozzi who have become regular companions in your life.

    Plus a wonderful mix of folk music, pop, blues and jazz.

    This special service that is public radio is made possible because of your ongoing financial support.

    So, on behalf of everyone here at the station, and your neighbors throughout North Texas, I want to thank you for renewing your membership for 1999 today.

           Sincerely,

           Jeff Luchsinger
           Station Manager

FIGURE 7.1 KERA 90.1 Membership Renewal Letter
    Source: KERA/KDTN, Dallas. Used with permission.

A section of the *1995 NACB Station Handbook* (National Association of College Broadcasters, 1995b) discusses grants within the context of fund-raising.

> This topic gets even more involved than on-air fundraising, but any station—TV or radio, cable or broadcast—has a shot if it qualifies. Despite the current economy, there is still considerable charitable giving being done by foundations, government agencies and corporations. Try to tap your school's development office for professional grant fundraising expertise as to how to write and present a grant proposal. ... Attempting to receive grants can require a great deal of knowledge of the fundraising system. One of the most difficult hurdles in applying for grants is finding a foundation that will likely donate money, as most foundations deal with larger organizations than college radio or television stations. (172)

However, the *NACB Handbook* goes on to say that "many local foundations may be interested in giving to a particular station in the area or to support a particular station program. The first challenge is to identify likely donors" (172). Chamber of commerce offices, along with libraries, offer a great source for searching out donors. Look for actual books and catalogs that reference foundations. Check with other stations that have been successful and discuss possible foundations as sources. Finally, "don't forget that approximately 80% of donations to non-profit organizations is by individuals" (172).

> From 1969 to 1978, public television and public radio experienced rapid growth on and off the campus. Two federal programs made vast sums of money available to civic and educational groups for the creation of local public broadcasting outlets. The first was PTFP or Public Telecommunications Funding Project; a grant program which provided money for the construction of station studios and transmitters. The second program, CPB or the Corporation for Public Broadcasting, provided for an annual grant which schools could use to defray operating expenses and to provide support for personnel (Carnegie Commission 1979). (Dennison 1994, 7)

Advice when writing your grant proposals: Follow the directions and rules precisely as outlined. Most grants require a written narrative at the beginning specifically describing your desire or project to be funded. Don't exceed the page limitation. Write clearly and state exactly what you want. Vague and inconsistent descriptions appear to be weak. Consider that obtaining the grant support may take some years, so be prepared to adjust your costs accordingly.

Underwriting Announcements

A major supporter of the day-to-day programming can come from underwriting. These are the "commercials" of noncommercial radio. There is a fine line between what can and cannot be said, with a great deal left to local interpretation by station management. Legal guidance is recommended if there is a question. The following provides a general overview of underwriting announcements.

The National Association of College Broadcasters' *1995 NACB Station Handbook* provides a thorough explanation of underwriting guidelines, which is included here:

UNDERWRITING ANNOUNCEMENTS

Underwriting announcements cannot encourage a listener to do something, like "apply for the credit card now," "for more information, call ..." or "come on down to the promotion at the new store opening." Underwriting announcements can include passive identifiers, like "credit card applications are available at the Student Union," "information is available by calling ..." or "the live promotion will take place at The Gap on Monday."

Raising money for a noncommercial organization is a multifaceted undertaking. Promotions and on-air fundraising are only two ways to raise money. Sponsorship and underwriting, objective forms of advertising, are the best ways to raise revenue for a college station. Underwriting is the type of advertising legal on non-commercial stations, and is best exemplified by sponsorship on PBS or NPR. In general, an underwriting spot does not make subjective statements about a product or promote sales. An example would be the comment: "Masterpiece Theater is made possible by a grant from the Mobil Corporation." Mobil benefits from name recognition, but the statement does not include any subjective comments concerning operations or products. This underwriting concept can be applied on a local level by any station, by appointing an underwriting director or staff and then approaching local businesses to sponsor shows. Local restaurants, clothing and record stores are often the most visible businesses on college campuses and the most interested in sponsorship on college television or radio stations.

Advertising on noncommercial broadcast college TV and radio stations occurs in the form of underwriting announcements. The underwriting rules get very confusing when it comes to promotional announcements bordering on commercial spots. The guidelines that follow are the parameters for legal underwriting announcements.

General Underwriting Guidelines

Underwriting announcements may take place at any natural break in programming (e.g., at the end of a program segment), including the beginning and end of a program.

Recommended underwriting spot length is 30 seconds maximum (although there is no legal restriction on announcement length).

Underwriting announcements generally begin with "This program is brought to you in part by..."

Do Not's

■ Underwriting announcements can identify, but cannot promote.

■ Underwriting announcements cannot include qualitative or comparative statements.

■ Underwriting announcements cannot include a call to action.

■ Underwriting announcements cannot mention discounts or savings percentage.

■ Background music cannot violate any of the above rules in a passive way that would be banned verbally.

Yes You Can

■ A sponsor's logogram can be contained in an underwriting announcement, so long as it doesn't conflict with other underwriting guidelines.

■ Still and moving images [for television] and other special effects may be used in underwriting spots as long as they do not violate the above rules or make the sponsor announcement appear too similar to a commercial.

Commercial Exceptions to the Noncommercial Rules

Many college stations think all their sponsor announcements have to be noncommercial. This is not necessarily true. Here are some legal exceptions allowing you to run full-fledged commercials:

■ If the announcement promotes a non-profit group or activity, you may run normal commercial spots (and charge for them accordingly). Of course, you may still choose to do PSAs [Public Service Announcements] for such events instead of, or in addition to, commercial spots.

■ During station-designated on-air fundraising campaign periods, you may solicit donations on behalf of your station with "self-commercials."

■ None of the underwriting rules apply if you are a carrier-current or cable radio station, or a closed-circuit or cable TV station, because the FCC's underwriting rules only relate to FCC-licensed broadcast stations.

However, if the policy of your station, school or cable system distributor says you must be non-commercial, remind them that there is no legal requirement to be that way and maybe you can get the policy changed.

Underwriting announcements cannot mention specific number percentage, like "10% off." Announcements must also avoid the concept of a discount, like "on sale now."

Thanks to the 1981 FCC ruling on enhanced underwriting, in addition to sponsor name, address and phone number, a "menu list" of items that the sponsor offers can be given.

An underwriting announcement can list products and services—though steer away from specific brand products like "Big Mac" or "Pontiac." However, the spot cannot "hype" the product or sponsor, e.g., "go buy these products."

A logogram is the motto which normally accompanies the sponsor's name in promotional and advertising materials. For example, "Waterford—Fine Crystal Since 1891," or "GE. We bring good things to life." as a trademark line, could be used in an underwriting spot.

But "Get Met. It pays." cannot be used in an underwriting announcement since it includes an implied call to action. [Note that the FCC has never ruled against a specific call to action in a slogan; therefore, this point has never been tested.]

Underwriting spots for video should not include "for sale" signs; no "great food" lyrics or jingles should be used behind audio spots.

Background music and video may be used that has no selling message, passive or overt.

Underwriting announcements should remain value-neutral. Flowery qualitative and comparative adjectives are prohibited, e.g., "sumptuous food," "best in town," or "fine crystal." (186-87)

See Figure 7.2 for a sample underwriting agreement form.

As far as Public Service Announcements are concerned, noncommercial stations, under current rules, are allowed to "sell time" for PSAs. As always, it is recommended that station management review up-to-date rules and regulations governing such.

Even trade-outs (cable companies, cellular phone services, local newspapers, and so forth) can work to the station's advantage. Product exchange is provided for "grant-in-kind" announcements. Additionally, underwriting may be considered a tax-deductible donation (highlighting the words *may be* in most cases). Determination of such should be made with caution and under advisement from the school's business office.

School development offices and athletic departments are great sources for names of potential supporters. This is where the school and station are truly working together to generate outside support.

Again, if there is any question regarding the legality of fund-raising efforts, consult legal counsel. Commercial broadcasters realize the influence noncommercial support can have on their dollars (see Cole 1998).

Finally, as mentioned in Chapter 3, the subject of political announcements needs to be raised, particularly regarding the realm of underwriting by candidates. First, bona fide candidates for federal office must be granted reasonable access to airtime at noncommercial stations. This does not necessarily mean "spot" announcements, but more in the line of airtime for commentary, discussion, and debate. The station can charge a fee to cover expenses incurred for such access, but cannot profit from it.

Noncommercial stations do not have to afford the same privilege to state or

RADIO STATION UNDERWRITING AGREEMENT

This Agreement is made this ____ day of _____, 1998 between:

(Radio Station Call Sign)	(Merchant)
Address	Address
Tele. / Fax Numbers	Tele. / Fax Numbers

Radio Station (Call Sign) ("the Station") is licensed to __(Community of License)__ and is authorized by the Federal Communications Commission ("FCC") to operate its noncommercial broadcast station on an assigned frequency of __(frequency)__. Pursuant to the rules and regulations of the FCC, the parties to this agreement acknowledge that the Station is permitted to broadcast underwriting and sponsorship announcements from for-profit entities, but is not permitted to broadcast commercial advertisements by for-profit entities. (In the event the above-referenced Merchant is a bona fide non-profit entity, the Station is permitted to broadcast commercial advertisements on behalf of the Merchant.)

Merchant desires to broadcast certain underwriting spots in conformity with this Agreement and all rules, regulations and policies of the FCC.

Broadcast Schedule: For value received, (Merchant) agrees to pay, and the Station agrees to broadcast underwriting announcements as follows:

Beginning Date: _____   Ending Date: _____

Specific Schedule Per Day/Week: _____

_____

Total Number of Announcements During This Contract Period: _____

Fees and Assumed Costs: _____ Dollars ($) per announcement during the term of this Agreement, payable in full, in advance, at the time this Agreement is executed.

Broadcast Announcements: Attached is the broadcast announcement(s) governed by this Agreement. The Station has final editorial control over the content of the announcement(s) and may revise, reject or terminate any such announcement(s) in order to maintain good faith compliance with relevant FCC rules and regulations.

Confirmation of Performance: At the conclusion of this contract term, the Station will provide Merchant with a written statement verifying the dates and times that each announcement was broadcast, including the total cost thereof.

Emergency Programming: The Radio Station retains the right to interrupt or preempt any announcement covered by this Agreement at any time in case of an emergency, or to broadcast other announcements or programs, if in its editorial discretion, to do so would best advance the Station's public interest responsibilities.

This Agreement shall be governed by, and construed and enforced in accordance with the laws of the State of __(state where the station is located)__.

This is the entire agreement of the parties in regard to these matters. There are no oral agreements existing between them.

The Station: **(Merchant's Name):**

By: _____   By: _____

Printed Name: _____   Printed Name _____

FIGURE 7.2 Radio Station Underwriting Agreement Form
Source: Booth, Freret, Imlay & Tepper, P.C. Used with permission.

local-office candidates. But if the station elects to do so, then it must make available airtime to other legitimate candidates. If the station does allow for candidates to underwrite programming, the announcement cannot encourage listeners to vote for the candidate, as this would violate underwriting rules and fall under the category of a commercial announcement. Again, consult legal counsel if questions arise.

Web-Site Advertising

Another form of advertising that noncommercial and commercial college radio stations are starting to take advantage of is that being done on the station's Web site. Some view this in the same realm as advertising sold in the station program guide, as discussed later in this chapter. An example from the Minnesota Public Radio Web site is provided in Figure 7.3. (See Peggy Miles's 1996 article, which provides examples of broadcast database and direct response programs on the Internet.)

FIGURE 7.3 "Welcome to Genway"
Source: Minnesota Public Radio. Used with permission.

In Chapter 9 and Appendix C I have included details about creating a Web site at the college radio station. Craig A. Stark also wrote in 1998 that, for the Web site to be successful, it should contain the following basic themes:

1. **General information about the station, department and university—** *Are there links to the station's department and university? Is a program schedule provided? Is there general information on the station or staff?*
2. **General information about the surrounding area the station serves—** *Can a user find local information? What kind of information is available?*
3. **Recruitment for the station, department or university—***Would a prospective college student want to join? Is there information on working with the station? Is the site appealing to a college student? Has the site been recently updated?*
4. **General reference materials that can be accessed by the user—***Are there any links that current students may be able to use to find reference material for courses? What types of references might they be?*

(Stark, 1998)

It is again recommended that any information on the station Web page always be checked for accuracy and be up-to-date. If the station management and/or programmer take the time and energy to set up a Web page, then it is incumbent upon them to work at running the site. How many Web sites are always "under construction" and never even truly started?

A list of noncommercial station Web sites is maintained by radio station KZSU at Stanford University and can be accessed via the Web at *http://kzsu.stanford.edu/other-radio.html*. This site will also direct you to another listing entitled the Gumbo Non-Commercial Radio list, which includes noncommercial stations without any Internet service. It can be located on the Web at: *http://www.gumbopages.com/other-radio.html*.

Additionally, in 1998 Steven McClung at the University of Tennessee in Knoxville was conducting a study for his doctoral dissertation about the uses of college radio Web sites, with an online survey based on a "uses and gratifications" theory. The results of his study were not available at the time of this writing (personal communication, October 5, 1998). Also, as a reference, see Glenn Gutmacher's list of radio station Web sites in Carl Tyrie's 1998 article "Building the Perfect Wave" (23, 27).

Newspaper Ads

Running your weekly program listing in the campus newspaper is a great way to promote your station. The newspaper can provide that tangible item for your listeners, just like the program guide. If you are designing your own ads to place in a newspaper, such as promoting a show or live broadcast, you will need to understand space sizes: sold based on columns for width and agate lines for depth (Longyear 1954, 49). Always consult the newspaper in which you are advertising for precise measurements. (See also Harvard Common Press 1978.)

If you advertise in your campus newspaper, why pay for it? Consider trading out underwriting or program guide advertising for newspaper advertising. Always be sure to double-check your newspaper ad before the paper run (when it's printed) for mistakes. Don't accept ads in the gutter (the center portion of the paper next to the fold), as ads placed in the outer portion of pages provide for better visibility. Let newspaper advertising complement your already on-air promotions.

The Station Program Guide

We're looking into getting some more advertisements in our Program Guide and were wondering what kind of rates other college stations are charging for space. ... We've been printing our Guide on newsprint—but are also wondering what other formats/styles college stations are using. (NACB ListServ, November 7, 1997)

As discussed in Chapter 3, a program guide can serve to promote station programming, and the station in general. Here the focus is on the program guide as a revenue-generating source, even to the point of paying for itself.

PROGRAM GUIDE

A program guide should be yet another extension of the station's positioning theme, a continuation of the station, in print. Varying from one page, to posters, to small magazines, the program guide will only be limited by imagination and funds. Selling packages that include print and air-time not only can fund the program guide, but also can gently introduce new clients to radio by using a medium with which they are familiar.

Setting advertising rates: Generally about half of the space in the program guide should be sold for advertising. Figure out how much each full page should cost to cover the full expense of producing the guide, and add in a percent of profit. The smaller space increments are then figured as a percentage— one half page should cost 75% of a full page, one quarter should cost 75% of a half, and so on. Make up a rate sheet for the sales staff to use and work together to develop packages with on-air time. (National Association of College Broadcasters 1995b, 203-4)

Another idea is to "spotlight" underwriters in the station program guide. At the least, a list of underwriters should be included.

Hundreds of years ago when I was in college we used to give the sponsors photos and a one-paragraph narrative when we gave away a prize that they donated. Delivering this usually prompted them to give us some other piece of slow-

moving inventory for the next promotion. (Airwaves Radio Journal Issue #2047, Airwaves Media ListServ, April 14, 1996)

Whether dealing with underwriting, program guide, Web site/Internet, or webcasting advertising, consideration should be given to paying the salesperson (student, volunteer, or staff member) a commission, in addition to any regular pay (either part- or full-time). The idea is that this is consistent with the real world and provides for proper compensation for bringing in actual dollar support for the station. Additionally, such "sales" experience can lead to a real job in the media industry by providing practical experience for the student.

Station Finances in General

I am preparing my budget for the 1996-97 fiscal year. I would be interested in how other managers prepare budgets, i.e., fund allocation for day-to-day operating expenses, travel, coverage of athletics, etc.

Without being too direct, what are the various operating budgets out there? (NACB Discussion List ListServ, February 27, 1996)

Station advisors and managers must be ready to prepare a station budget proposal annually. The academic department in which the station operates should be able to assist in the setup. Utilities (electric, water, natural gas, etc.) and building maintenance (upkeep and custodial) costs are usually provided by the school and not included in the proposal. Be sure to include required matching with salaries (full-time, part-time, and contract). Such items include any health insurance, Social Security, FICA/Medicare, retirement, federal and local matching taxes, and administrative overhead. Concerning the salary for the chief engineer (a position described in Chapter 6), if this individual also serves as the academic department audio or video engineer, the department should consider paying his or her wages (or at least splitting them with the station). Sample line items included in a budget are provided in Figure 7.4. Note that this is an annual budget *after* the station has been constructed.

It is recommended that managers consult with such groups as the National Association of College Broadcasters and the Intercollegiate Broadcasting System for guidance in preparing station budgets. Different types of stations (broadcast versus nonlicensed, NPR affiliates, and so on) have different needs.

Managers and advisors must learn to appreciate unexpected costs. Transmitter tubes and new equipment requirements can be costly. If possible, try to have a contingency fund available for unforeseen expenditures.

New equipment is sometimes the hardest to be funded. At times, requesting a piece of equipment a few years in a row is necessary to convince the funding authorities that it is really needed. Always consider outside donations for allocation to new equipment—possibly as a station campaign or fund-raiser.

Also, consideration should be given to two other sources of station revenue

Salaries and Wages
      Noncontract (Full/Part Time)
      Contract
Maintenance and Operation
      Wire Service (AP, UPI)
      Computer Costs for Wire Service
      Audio News Services
      Program/Satellite Distribution Services
      Subscriptions and Special Publications
      Telephone (Studio, Remote, Transmitter Site)
      Long Distance Phone Charges (including cellular)
      Printing and Copying
      Office Supplies
      Recording Supplies
      Studio Facilities Maintenance—Minor
      Transmitter Facilities Maintenance—Minor
      Tower Maintenance—Minor
      Programming Supplies
      Postage
      Postal Service Box Rental
      Music Licenses (ASCAP, BMI, SESAC)
Travel
      Management/Programming
      Student Participation
          Sports
          Broadcast Conferences
Major Repairs and Rehabilitation
Capital Outlay

FIGURE 7.4 Sample Station Line-Item Budget

generation: tower space rental and SCA (subcarrier allocation) leases. If the station (really the school) owns the tower the station operates from, the station can lease tower space to other stations, paging services, and cellular telephone operators (see Rusk 1996b). Additionally, you can lease your subcarrier frequency (see Rusk 1996a). This is one area where the nonprofit station can generate funds directly from for-profit businesses. Engineering consultation will probably be necessary to determine market coverage potential of the SCA frequency and to ensure that whatever goes on the tower does not affect the radio station's transmission. Additionally, it is recommended that one "shop around" to determine going market rates for such leases so that the station doesn't underprice itself.

Whatever revenue is generated from the outside by the station should be delegated to ensure that the money goes to the station itself. In 1998 Yana Davis reported that "[m]any university licensees are now being asked to generate surplus (revenues) and return it to the universities" (Yana Davis 1998, 17). This is where working with the school administration to communicate the college radio station goals and purposes can pay off.

Funding Auxiliary Enterprises

The subject of college radio stations operating in the context of auxiliary enterprises is examined in Chapter 2. Here, the same issue is examined as it relates to college radio station funding. Almost all of the previous research pertaining to national budgeting and funding patterns relating to college and university auxiliary enterprises has centered on accounting procedures. This research has made clear that the application of auxiliary enterprises normally applies to for-profit units. Here that research is described in detail in the context of how it can relate to funding a college radio station as an auxiliary enterprise. The reader is once again directed to my own major research in my doctoral studies, culminating in my dissertation, *An Analysis of Selected Factors Which Influence the Funding of College and University Noncommercial Radio Stations as Perceived by Station Directors* (Sauls 1993), as this work focuses entirely on college radio station funding.

In a manual designed to assist college business officers in accounting systems and management reports, Hughes (1980) points out that "a college or university is a complex organization composed of many units designed to accomplish different purposes or functions" (55). Also noted by Hughes, these groups are classified into units such as instruction, academic support, student services, and auxiliary enterprises, depending upon the function they perform. In specifically addressing auxiliary enterprises, Hughes provides the following description: "An auxiliary enterprise is an entity that exists to furnish goods or services to students, faculty, or staff, and that charges a fee directly related to, although not necessarily equal to, the costs of the goods or services. The distinguishing characteristic of auxiliary enterprises is that they are managed as essentially self-supporting activities" (96). Hughes further distinguishes between two types of auxiliary enterprises: those that primarily furnish services to students, and those that provide services to faculty and/or staff (96).

Units were described further in an earlier publication by Mertins and Brandt (1970) entitled *Financial Statistics of Institutions of Higher Education: Current Funds Revenues and Expenditures 1968–69.* As a result of data obtained from the fourth annual Higher Education General Information Survey, revenues and expenditures were listed in the following categories: (a) educational and general revenues (expenditures), (b) student aid grants, (c) major public service programs, and (d) auxiliary enterprises.

In a 1969 study sponsored by the Ohio House of Representatives, management in Ohio public higher education was addressed:

> Auxiliary Enterprises are service operations conducted to the benefit of students and faculty. Since these enterprises are completely controlled and funded by the individual university, their expansion or curtailment does not require state approval nor are state funds made available for these purposes. Fees are charged for these services, and they are intended to be self-supporting.
>
> At Ohio's public universities, the major auxiliary enterprises—in terms of monies expended—are:

- Residence and dining halls
- Student unions
- Bookstores
- Intercollegiate athletics
- Student health services (57)

The report is unique in that it specifically addresses media equipment in auxiliary enterprises. Although the report recommends that "the academic resources of the library, university computer services, audio-visual resources, and television should be combined into a single division" (57), it appears that the use of media equipment is specific to that of an academic nature, rather than to broadcasting.

In dealing with funding specific to auxiliary enterprises, Abel (1982) notes in a report entitled "Comparative Information on Higher Education," that "almost half of total current funds revenues of public institutions in the SREB [Southern Regional Education Board] states comes from state appropriations: tuition and fees, federal government contracts and grants, and auxiliary enterprises each provide around 11 percent" (4). Abel's study addresses the application of funds generated by auxiliary enterprises.

Bock and Sullins (1987) address alternative sources of funding in community colleges and private fund-raising in order to show that auxiliary enterprises can be funded through nontraditional methods. They note that "a college foundation can ensure the survival of a special project that might otherwise go unfunded by rallying community support to generate the required revenue" (14). Bock and Sullins suggest that such avenues as life income gifts, annuity programs, and corporate solicitation be investigated as possible revenue sources (15-16). An additional idea presented in their study was that, "as an alternative or additional resource, community college leaders ... commit themselves to commercial pursuits through expanded auxiliary enterprises or revenue diversification" (16). They note that "some have adopted the perspective that the private solicitation of resources by public community colleges result in unhealthy competition for the financially strapped private institutions" (18), but added that "attempts by community colleges to gather supplemental private resources, while a commentary on a public policy of fiscal restraint, should never be interpreted as intended to relieve the states of their financial responsibility for educating this nation's citizenry" (18).

Community colleges and auxiliary enterprises are also addressed by Stumph (1985) in an article in the journal *New Directions for Community Colleges.* Stumph states that "the National Association of College Auxiliary Services provides a very simple, straightforward definition: 'Auxiliary services is that division of a college whose operations furnish a variety of goods and services for the support of the institution's education program'" (Clark, cited in Stumph 1985, 65). Stumph interprets auxiliary enterprises, most commonly bookstores and food services, as noneducational services under a broad framework. Stumph adds that auxiliary enterprises have the ability to be "free from operating and financial restrictions that may be imposed by law and personnel regulations that are attuned to the academic rather than business world" (68). Advantages of auxiliary enterprises, particularly for community colleges, have historically been "to provide a necessary service and convenience to students, staff and visitors. An additional incentive has been to provide opportunities for student employment on campus" (71).

In addressing sources of funds for campus capital renewal and replacement, Kaiser (1984) presents the following explanation: "Sometimes profitable and at other times requiring subsidies, educational activities and auxiliary enterprises are mixtures of moments of inspiration and implied obligations to faculty, staff, students and public. ... In some cases they are necessary services and in other cases, luxuries" (10).

Young and Geason (1982), in a paper entitled *Cost Analysis and Overhead Charges at a Major Research University,* describe a cost allocation model at Ohio State University designed to measure the direct, indirect, and total operating costs of university operations (in particular those associated with unrestricted general fund support of auxiliary enterprises and revenue-generating activities). They report that "the set of auxiliary enterprises and earnings operations at a major research university is incredibly diverse in terms of clients served and operational scope" (5).

In a 1986 article published in the *Chronicle of Higher Education,* Heller addresses the use of auxiliary enterprises. As a result of a report prepared for the National Association of College Stores, Heller explains that, "though the report offers no prescriptions, it outlines four alternatives to institutional control or leased management, which are now the most common arrangements for bookstores and food services" (21). Those four alternatives are categorized as cooperative, for-profit subsidiary, nonprofit subsidiary, and foundation. Heller also notes that some small-business operators have charged colleges and other nonprofit organizations with unfair competition in the use of their auxiliary enterprises (21).

Welzenbach, in a 1982 book, *College and University Business Administration* (cited earlier in Chapter 2), addresses the college or university's role in funding auxiliary enterprises: "Senior administrators should insure that policies supporting an institution's desired style and quality of life are understood, and that managers of enterprises adopt and implement these policies. Auxiliary enterprises are recognized vehicles for attracting and retaining students, faculty, and staff" (198).

In a 1990 dissertation study focusing on the patterns of higher education funding in the state of Michigan, Nicosia comments that "the general public may incidently be served in some auxiliary enterprises" (184). Nicosia also observes that "the more affluent institutions tend ... to spend a greater percentage on auxiliary enterprises than the less affluent institutions" (189).

Hogarth (1987) addresses auxiliary enterprises as part of support services and activities, in his book entitled *Quality Control in Higher Education.* His discussion of the topic is limited, however, to the areas of food services, student housing, and campus security (104-8).

Although all of the related research reviewed in this section, and in Chapter 2, addresses how colleges or universities might consider a noncommercial radio station as an auxiliary enterprise, no reports on national studies of college and university noncommercial radio as an auxiliary enterprise were found. Thus, no research concerning the funding of such an enterprise was available.

Noncommercial or Commercial?

Finally, proponents for a college radio station must address the choice of applying for a noncommercial or commercial license. Charles Bailey (1993) sums up the dilemma faced by administrators and managers:

> Advocates for a campus radio station should consider the favorable and unfavorable features of a commercial and a noncommercial radio station. Holgate implies that there is more prestige with a commercial station, employing full-time professionals and using a few students in minor station positions. In contrast to a commercial station, a noncommercial campus station will be student staffed and seems to be accepted by the local listening audience as "training grounds" for students (Holgate 1982). Also, administrators should determine if

a campus commercial station "can compete for revenues and audiences with local commercial stations, has the support of local and statewide broadcasters, community, and campus, and can be self-sustaining or at least pay for a professional staff" (Holgate 1982, 6). Since noncommercial stations are non-profit, Holgate says that a campus noncommercial radio station will not compete with the commercial stations in its listening area. (106)

As mentioned in Chapter 2, many schools operate more than one type of station. Some schools operate both noncommercial and commercial outlets. For example, the University of Florida operates the following stations:

WRUF-AM 850 Kh[z], commercial news/talk ("Radio 850")
WRUF-FM 103.9, commercial AOR ("Rock 104")
WUFT-FM 88.9 MHz, noncommercial NPR affiliate ("Classic 89")
WJUF-FM 90.1 MHz, duplicates the programming of WUFT-FM for the area
  around Citrus County ("Nature Coast 90")
WUFT-TV channel 5, Cox Cable/University City channel 11, PBS member station ("Where Learning Never Ends")
WLUF-LP channel 10, Cox Cable/University City channel 6
Radio Reading Service for the Blind, subcarrier on WUFT-FM
There is a gateway to all of these stations at http://www.jou.ufl.edu/
  stations.htm. The commercial stations' Web sites actually have streaming
  audio and video. (Personal communication, September 23, 1998)

Additionally, public radio stations, in particular, will utilize translators and repeater stations to extend their broadcast service.

REFERENCES

Abel, R. L. (1982). *Comparative information on higher education.* Atlanta: Southern Regional Education Board.
Bailey, C. G. (1993). Perceptions of professional radio station managers of the training and experience of potential employees who have worked in college radio under one of three different administrative patterns. Ph.D. diss., West Virginia University, 1993. Abstract in *Dissertation Abstracts International,* A54/08:2809.
Bock, D. E., and W. R. Sullins. (1987). The search for alternative sources of funding: Community colleges and private fund-raising. *Community College Review* 15(3): 13-20.
Brant, B. G. (1981). *The college radio handbook.* Blue Ridge Summit, Pa.: TAB.
Carnegie Commission. (1979). *A public trust: The report of the Carnegie Commission on the future of public broadcasting.* New York: Carnegie Corporation of New York.
Clinton's 1997 budget plan for higher education and science. (1996). *Chronicle of Higher Education 47(29):* A42-43.
Cole, H. (1998). That's enough of that funky stuff: Are noncommercial stations crossing the line with sponsor announcements? *Radio World* 22(13): 27, 30.

Conciatore, J. (1995). The trade-offs of public radio. *Radio World Magazine* 2(12): 17-23.

Corporation for Public Broadcasting. (1995-96). *Telecommunications for the public interest.* Washington, D.C.: Corporation for Public Broadcasting.

———. (1996). *Public broadcasting's service for the American people.* Washington, D.C.: Corporation of Public Broadcasting.

Dennison, C. F., III. (1994). Administrative patterns of on-campus radio stations and the leadership behaviors of the managers. Ph.D. diss., West Virginia University, 1994. Abstract in *Dissertation Abstracts International* 54/11:3940A.

Evans, G. (1986, January 22). Howard U.'s radio station rated no. 1 in highly competitive Washington area. *Chronicle of Higher Education* 31(19): 3.

Gronau, K. (1995). New artists, public broadcasting, NAC regenerate traditional jazz. *Radio World Magazine* 2(11): 20, 22.

Harvard Common Press. (1978). *How to produce a small newspaper.* Harvard, Mass.: Harvard Common Press.

Heller, S. (1986). New legal structures suggested for colleges' auxiliary enterprises. *Chronicle of Higher Education* 33(11): 21.

Hogarth, C. P. (1987). *Quality control in higher education.* Lanham, Md.: University Press of America.

Holgate, J. F. (1982). Determining the role of campus radio. *Journal of College Radio* 19(2): 4, 6-7.

Hughes, K. S. (1980). *A management reporting manual for colleges: A system of reporting and accounting.* Washington, D.C.: National Association of College and University Business Officers.

Kaiser, H. H. (1984). *How can we afford this? Funding and financing means.* ERIC, ED 252 148.

Knopper, S. (1994). College radio suffers growing pains. *Billboard* 106(28): 84.

Landay, J. M. (1996). Don't make public television stations commercial. *Christian Science Monitor,* July 22.

Longyear, W. (1954). *Advertising layout.* New York: Ronald Press.

Lucoff, M. (1979). The university and public radio: Who's in charge? *Public Telecommunications Review* 7(5): 2-26.

Mertins, P. F., and N. J. Brandt. (1970). *Financial statistics of institutions of higher education: Current funds, revenues and expenditures 1968-69.* Washington, D.C.: National Center for Educational Statistics.

Miles, P. (1996). The Internet: Future of direct marketing for radio. *Tuned In* 3(6): 41-42.

National Association of College Broadcasters. (1995a). *1995 college radio survey.* Providence, R.I.: National Association of College Broadcasters.

———. (1995b). *1995 NACB station handbook.* Providence, R.I.: National Association of College Broadcasters.

National Association of Educational Broadcasters. (1967). *The hidden medium: A status report on educational radio in the United States.* New York: Herman L. Land Associates. ERIC, ED 025 151.

Nicosia, P. C. (1990). The patterns of higher education funding in Michigan: Its implications for institutional resource allocation. Vol. 1-2. Ph.D. diss., University of Michigan, 1990. Abstract in *Dissertation Abstracts International,* 52/01A:91.

Ohio House of Representatives. (1969). *Management study and analysis. Ohio public higher education.* Chicago: Warren King. ERIC, ED 031 152.

Petrozzello, D. (1995). Public radio's worst fear: "Zero funding means death." *Broadcasting and Cable,* March 6, 50.R

Rusk, B. (1996a). Legal advice on subcarrier leasing. *Radio World* 20(13): 9.

———. (1996b). Making money multiply on towers. *Radio World* 20(14): 7, 11.

Sauls, S. J. (1993). An analysis of selected factors which influence the funding of college and university noncommercial radio stations as perceived by station directors. Ph.D. diss., University of North Texas, 1993. Abstract in *Dissertation Abstracts International* 54:4372

———. (1996). College Radio: Points of contention and harmony from the management perspective. *Feedback* 37(2): 20-22.

———. (1998). Factors that influence the funding of college and university media outlets: Radio as a blueprint. *JACA, Journal of the Association for Communication Administration* 27(3): 163-71.

Stark, C. A. (1998). *What your campus radio station Web site needs to be successful.* Paper presented at the Texas Association of Broadcast Educators Conference, Dallas, September 4, 1998.

Stumph, W. J. (1985). Auxiliary and service enterprises. *New Directions for Community Colleges* 13(2): 65-71.

Tyrie, C. (1998). Building the perfect wave. *College Broadcaster* 11(2): 22-23, 27.

Welzenbach, L. F., ed. (1982). *College and university business administration.* 4th ed. Washington, D.C.: National Association of College and University Business Officers.

Yana Davis, S. D. (1998). Non-coms trade marketing savvy. *Radio World* 22(17): 17.

Young, M. E., and R. W. Geason. (1982, May). *Cost analysis and overhead charges at a major research university.* Paper presented at the annual forum of the Association for Institutional Research, Denver, Colo. ERIC, ED 220 035.

8

The College Radio Station and the Community

In April 1995 Scott Frampton, editor in chief of *CMJ New Music Monthly*, a college radio trade magazine, stated: "College radio is providing a service to the community, providing programming you can't get anywhere else on the dial. It should be more than just the campus jukebox" (McDonald 1995, 21). Regardless of its transmission mode (as outlined in Chapter 2), the college radio station will inevitably have an impact on the community, both on and off campus, that it serves.

The National Association of College Broadcasters has described the college radio station's public file thusly:

> This list should summarize those issues facing its community (primarily the city of the license and only secondarily for the greater service area) to which the licensee paid particular attention with programming during the previous three months, together with a brief description of how each issue was addressed by the licensee. In addition, a list of the licensee's most significant programs responsive to each should be provided, ... together with a brief description of the program, demonstrating how it significantly addressed the issue and whether or not it was locally originated. (1995, 94)

In reality, for some college towns, a majority of the issues facing the community in general may emanate from the campus from which the college radio station broadcasts. In this sense the campus community indeed does become part of the local community.

The college radio station can also be a community radio station in the truest sense. It is not unusual for the campus radio to use community citizens as station volunteers. Proceeding from the premise that "community radio is about and for the community," one would expect that people are on the air because they have something to say or share. Thus, though they may not be the most professional in their delivery, their purpose is sincere.

THE CAMPUS COMMUNITY

Dilemma: The campus community continually makes demands on the campus station for coverage, but lacks knowledge of the staffing, funding, and operation limitations of the station itself.

Not to be confused with the local community, the campus community can have great expectations of what the campus station should be doing. These expectations exist within administration, athletics, academic units, faculty, and, of course, students. Everyone on campus has a justifiable vision of what the campus radio station should be doing. What everyone does not realize is that resources (financial, physical, and staff) are limited. Additionally, it is the intrinsic responsibility of the radio station itself, under the direction of its management, to determine its own purpose. "As a result, it is [sometimes] necessary to choose certain types of programs as more desirable than others" (WSRN, 1991). Outside of direct programming control by the school and/or an academic unit, the station itself must determine its direction and provide the necessary continuity to carry out its stated mission. It should then communicate its purpose to the campus and the local community.

The ongoing broadcasts provided by college radio help to serve as public relations arms for the schools themselves. Often college radio stations are the only outlets for such broadcasts as campus sports and news. In regard to the colleges' and universities' perceptions of college radio, one advantage is that the institutional image is enhanced every time a well-programmed station identifies itself as affiliated with the school (Sauls 1995). Here the school should credit the station. But at times, for whatever reason, many campus stations go unrecognized. For example, the campus station may provide complete coverage of football games while the local commercial radio outlet will give only limited coverage. But the school will fully credit the commercial station for its efforts, while not recognizing the campus station. This tends to give the campus station staff (usually students) a negative impression of their own efforts (see Sauls 1996).

On a more positive note, the college radio station can utilize the campus community by turning to the faculty and staff as experts. This can be particularly helpful in the coverage of news events. Many campus PR or public information offices maintain such lists pertaining to areas of expertise and faculty/staff contacts.

THE LOCAL COMMUNITY

It is postulated that the local community is "confused" concerning the station.

Overall, as with commercial stations, the underlying premise of the college radio station is to serve the community, whether it be the campus community or the community at large, but in unique ways often geared to underserved niches of the population. This ideal is consistent with the fact that colleges and universities are licensed to "operate broadcast facilities in the public interest, convenience, and necessity" (Ozier 1978, 34). Studies indicate that the service component to the community is important (see Sauls 1993). And, of course, one cannot overlook the overall economic impact the college or university has on the local community in general (see Stokes and Coomes 1996).

"Nearly all stations see their primary function as one of providing alternative programming to their listening audiences. ... More specifically, the alternative programming is primarily made up of three types: entertainment, information, and instruction" (Caton 1979, 9). "College radio is as varied as college towns or college students" (Pareles 1987, C18). Some stations mirror commercial radio, while others opt to develop their own style.

Because the community does not understand the operation of the station, recognition of the station is sometimes not properly given. Additionally, demands put on the station are sometimes unrealistic or misleading, since the community does not know how the station functions. Broadcast requests made by outside sources are often beyond the scope of the intended purpose of the campus station (see Sauls 1996). For example, outside of an "entertainment calendar," which provides listings of upcoming events, noncommercial stations are prohibited from airing promotional announcements for a concert sponsored by a for-profit entity. Concert promos for nonprofit organizations are acceptable. Legal points such as these are often difficult for the general public to comprehend.

And the college radio station needs to understand its responsibility toward the community. In February 1998 the Associated Press reported that "[a] college radio station suspended live, on-air dedications after police said young listeners were using the broadcasts to promote gang activity." Another example of the station's ongoing commitment to serve the community is the idea of "equal time," already reviewed in Chapter 3.

One result of the Telecommunications Act of 1996 was the virtual elimination of the threat of "comparative renewal" challenges by competing applicants at renewal time, brought on by a congressionally endorsed two-step renewal process. What this means is that, as far as the FCC is concerned, nonentertainment programming (news, public affairs, public service announcements, and so on) is no longer required (see Cole 1996). As such, noncommercial stations, which tend to program these types of programs in prime time, will be taking up the slack left by commercial broadcasters who abandon such programming. In the end, particularly for the listener, this can be a good thing, with the growing presence of nonentertainment programs on college radio stations. Furthermore, the noncommercial station will look even better in the public's view for providing such programming. For the local community, here lies the true benefit of college broadcasting.

Again, the community often confuses the campus station with the commercial stations in the area. This can lead to crediting the wrong station. As always, the college radio station staff can never forget that their broadcast signal goes far beyond the campus boundaries. And even if the signal is campus only (such as those under part 15 of the FCC Rules as outlined in Chapter 2), that signal is still serving a community.

THE CAMPUS RADIO STATION VERSUS
THE CAMPUS NEWSPAPER

The campus newspaper is just that—a "campus" outlet. The campus newspaper does not carry the obligations toward community that the campus radio station carries. Though more visible than the radio station, it has no ongoing commitment to the local community, so it can stop printing, for instance, during school break periods. The campus radio station, however, is a federally licensed agent of the school and must meet minimum obligations so as not to jeopardize its continued authorization to operate.

Interestingly, the campus newspaper is often referred to as the "student newspaper," whereas the campus radio station is referred to as the "school station." So there is an opposition, then, between "student" and "school"?

DEREGULATION—THE SELLING OF COLLEGE RADIO

As was noted in Chapter 7, the few commercial stations licensed to educational institutions are able to seek potential advertisers. Some of these stations, such as Howard University's WHUR-FM, have even been successful in obtaining superiority ratings in major markets (Evans 1986). Thus begins the journey toward true commercialization of college radio.

Even more pressing is the entire deregulation of the broadcasting industry, and its possible impact on college radio. Already addressed in Chapter 7 is the threat of cuts to the funding for the Corporation for Public Broadcasting. F. Leslie Smith, coauthor of *Perspectives on Radio and Television: Telecommunications in the United States* (Smith, Wright, and Ostroff 1998), advises that when addressing the question "What deregulation?" one should refer to the "Summary of Deregulatory Policy Changes" presented by John W. Wright II in their 1995 book (Smith, Meeske, and Wright, 260-65). Smith notes *"that some of those are areas in which broadcasters used to find themselves in trouble quite frequently—e.g., nonentertainment programming guidelines, formal ascertainment procedures, and limits on commercials"* (Airwaves Radio Journal ListServ, Issue 837, September 28, 1994). Of course, many of these areas are applicable to noncommercial educational stations.

What truly is the role of the government's deregulation of the broadcasting industry? In the February 7, 1996, issue of *Radio World,* Lucia Cobo wrote:

The issues that the FCC controls number many more than those which directly affect broadcasters. Aside from protecting stations from each other, from technical interferers and pirates, etc., the commission can best be relied on to nurture new technologies as the only agent with its sole purpose in life to regulate that kind of business and the management of our spectrum.

The move to deregulation has been well-intentioned. Congress, however, should move toward it with prudence, mindful of those it could hurt with its legislation. (4)

For us in college radio broadcasting, deregulation has been felt most predominantly in the area of day-to-day oversight of station operation within legal realms. Initially, when the FCC dropped the requirement for operators to be tested to obtain the restricted radio-telephone operator permit (known as the third-class license), the burden of ensuring that DJs knew what they were doing was placed on the shoulders of the stations themselves, instead of on the individual. (Information pertaining to radio operators' licenses may be obtained via the Federal Communications Commission Audio Services Division Web site at *http://www.fcc.gov/mmb/asd/main/other.html.*) Then, most recently, the FCC began allowing for unattended operation, not even requiring a licensed operator on site. Although both of these changes relieved operators and stations of some governmental oversight, stations are still required to operate within legal parameters. Training takes on a whole new meaning.

Most troublesome, though, is the aspect of the selling of college radio brought on by deregulation and its effects on the broadcasting industry in general. Deregulation has made broadcasting big business (see Shapiro 1997)—so much so that college radio stations are becoming very lucrative as assets for sale on the campus. Case in point: It was reported by the Intercollegiate Broadcasting System in 1997 that a university radio station was slated to be sold to a religious-affiliated organization for a reported $13 million price tag. The university, it was said, put the station up for sale in order to "close a severe budget gap." The IBS wrote, "this kind of sale is just another reminder of the value of the frequency your station utilizes" (Intercollegiate Broadcasting System 1997). Although the station was eventually sold to C-SPAN (see Wiley 1997), this anecdote does highlight the fact that the religious groups are rigorous in obtaining frequencies, including noncommercial allocations held by colleges and universities. And the fact that these groups can afford large sums of money for the acquisition of such channels weighs heavily on its impact for the future of college radio. Schools will take into serious consideration such offers. It is noted that as colleges and universities are recognizing the dollar value of their stations and relinquishing such properties for an influx of "cash," they give up these stations forever as new frequencies in many markets are literally no longer available within the assigned broadcast spectrum of today.

Additionally, at the time of this writing, the FCC was seriously considering the idea of "auctioning" radio frequencies within the broadcast band. How this could or will effect the noncommercial allocations is yet unknown. But one take on the subject is particularly evocative:

> Here's a test. What exists in nature, costs nothing to produce, is owned by the government, is extremely rare and valuable, and can help balance a budget if auctioned to the highest bidder? Oil exploration rights in Alaska? Give up? You guessed it: radio spectrum. (Montero 1996, 19)

As was mentioned in Chapter 2, stations in the noncommercial portion of the spectrum are not included in the auction rules, but comments have been solicited "on whether noncommercial applicants competing for licenses on the commercial FM band should participate in auctions as well" (Stimson 1998, 14).

PRACTICAL APPLICATIONS

Contacts and Information Regarding the FCC

Because so many issues concerning the relationship between the college radio station and the community, be it on or off campus, also concern federal rules and regulations, it is of utmost importance that station management and engineering keep up with FCC mandates. The following three sources are thus provided:

1. [The] **Federal Communications Commission** provides through the World-Wide Web information about telecommunications and other areas. Users can retrieve information on F.C.C. licensing fees, a listing of the commission's staff members, and the *Daily Digest,* a synopsis of commission orders, news releases, public notices, and speeches by commissioners: *http://www.fcc.gov/.* (Floyd 1996, A22)

2. A great companion to the FCC site is *www.radiostation.com*, operated by Elliott Broadcast Services. The centerpiece of the site is its FM and AM databases. Again, you can search by many criteria and pull up useful data. What makes this website valuable is its ability to generate maps. The database can draw a downloadable map showing the transmitter site of the station you punch in.
   The site also includes FCC daily updates of actions and applications. (Somich 1997, 54)

(A sample map from *www.radiostation.com* is shown as Figure 8.1. These maps are generated from a server at the U.S. Census Bureau. Their Web-site address is *http://tiger.census.gov.*)

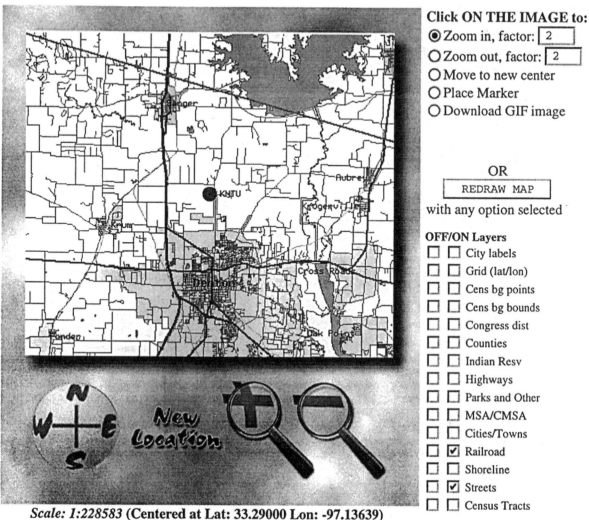

Scale: 1:228583 (Centered at Lat: 33.29000 Lon: -97.13639)

Click ON THE IMAGE to:
- ⦿ Zoom in, factor: `2`
- ○ Zoom out, factor: `2`
- ○ Move to new center
- ○ Place Marker
- ○ Download GIF image

OR

| REDRAW MAP |

with any option selected

OFF/ON Layers
- ☐ ☐ City labels
- ☐ ☐ Grid (lat/lon)
- ☐ ☐ Cens bg points
- ☐ ☐ Cens bg bounds
- ☐ ☐ Congress dist
- ☐ ☐ Counties
- ☐ ☐ Indian Resv
- ☐ ☐ Highways
- ☐ ☐ Parks and Other
- ☐ ☐ MSA/CMSA
- ☐ ☐ Cities/Towns
- ☐ ☑ Railroad
- ☐ ☐ Shoreline
- ☐ ☑ Streets
- ☐ ☐ Census Tracts

| REDRAW MAP |

If your browser doesn't support client-side imagemaps, use the controls below to navigate the map.

	NW	N	NE	
Zoom In	W	Pan	E	Zoom Out
	SW	S	SE	

NEW! 3D Mapping

Here is the FAQ and instructions on how to include these maps in your own web documents. The old mapbrowser has been moved to a new location.

FIGURE 8.1 Sample Downloadable Transmitter Site Map

3. Finally, an infobase providing an up-to-date copy of the Federal Communications Commission Rules is available from the Rules Service Company, 7615 Standish Place, Rockville, MD 20855, (301) 424-9402, FAX: (301) 762-7853. Stations will need to select which parts of the rules to acquire. Recommended parts are Part 17 Construction, Marking, and Lighting of Antenna Structures; Part 73 Radio Broadcast Services; and Part 74 Experimental, Auxiliary, and Special Broadcast and Other Program Distributional Services.

Of course, throughout this book many Web sites are suggested for those desiring more knowledge about college radio. These sites can help you locate needed information and provide phone/fax numbers and e-mail addresses for individual contacts.

REFERENCES

Associated Press. (1998). Radio station ends live dedications after police warn of gang use. Associated Press: February 23.

Caton, B. (1979). Public radio in Virginia. Virginia State Telecommunications Study Commission, Working Paper No. 12, Richmond. ERIC, ED 183 209.

Cobo, L. (1996). Deregulation is not radio's panacea. *Radio World* 20(3): 4.

Cole, H. (1996). Farewell to public service programming. *Tuned In* 3(5): 30-31.

Evans, G. (1986). Howard U.'s radio station rated no. 1 in highly competitive Washington area. *Chronicle of Higher Education* 31(19): 3.

Floyd, B. P. (1996). Information technology resources. *Chronicle of Higher Education* 47(29): A22.

Intercollegiate Broadcasting System. (1997). WDCU-FM being sold for $13 million. *IBS Radio Newsletter,* September/October.

McDonald, Glenn. (1995). Left of the dial. *U. Magazine,* April, pp. 20-21.

Montero, F. (1996). Going once, going twice, $old! *Tuned In* 3(2): 19-20.

National Association of College Broadcasters. (1995). *1995 NACB station handbook.* Providence, R.I.: National Association of College Broadcasters.

Ozier, L. W. (1978). University broadcast licensees: Rx for progress. *Public Telecommunications Review* 6(5): 33-39.

Pareles, J. (1987). College radio, new outlet for the newest music. *New York Times,* December 29.

Sauls, S. J. (1993). An analysis of selected factors which influence the funding of college and university radio stations as perceived by station directors. Ph.D. diss., University of North Texas, 1993. Abstract in *Dissertation Abstracts International* 54:4372.

———. (1995). College radio. Entry submitted to the *Encyclopedia of United States Popular Culture.* Bowling Green, OH: Popular Press. Santa Barbara, Calif.: ABC-Clio, forthcoming.

———. (1996). College radio: Points of contention and harmony from the management perspective. *Feedback* 37(2): 20-22.

Shapiro, E. (1997). Golden oldie: A wave of buyouts had radio industry beaming with success. *Wall Street Journal,* September 18.

Smith, F. L., M. D. Meeske, and J. W. Wright II. (1995). *Electronic media and government: The regulation of wireless and wired mass communication in the United States.* White Plains, N.Y.: Longman.

Smith, F. L., J. W. Wright II, and D. H. Ostroff. (1998). *Perspectives on radio and television: Telecommunications in the United States.* 4th ed. Mahwah, N.J.: Lawrence Erlbaum.

Somich, J. (1997). Code of federal regulations. *Radio World* 21(18): 54.

Stimson, L. (1998). License auction rules set. *Radio World* 22(18): cover, 14.

Stokes, K., and P. Coomes. (1996). The local impact of higher education. Paper presented at the annual meeting of the Association for Institutional Research, Albuquerque, May 1996. ERIC, ED 397 738.

Wiley, Ed, III. (1997). Short-circuiting college airwaves. *Black Issues in Higher Education* 14(15): 16-19.

WSRN. (1991). *WSRN programming philosophy.* Swarthmore, Pa.: WSRN-FM.

9

The Future of College Radio

N ow we look to the future of college radio. This, in itself, is difficult under any circumstance to make predictions about. But at least futuristic "expectations" can be proposed, even if we can safely assume they may become dated very quickly. This is particularly true when addressing such issues as technology and systems applications (as in this chapter), because these factors can become outdated very quickly within a limited life span. It is important for the reader to take note of this date-sensitive aspect. A good example of an attempt at predicting the future of radio broadcasting is the November/December 1998 issue of *BE Radio* magazine, entitled "The Future for Radio." Included is a compilation of issues dealing with technical, programming, and product innovations anticipated in years to come.

Aspects that deal with such issues as motivating students and changing attitudes in regard to the future of college radio broadcasting are found in the final chapter of this book. Program content issues are addressed in Chapters 3 and 4.

NONCOMMERCIAL VERSUS COMMERCIAL

How, then, does noncommercial college radio compare to commercial radio broadcasting? College radio "is pretty much unregulated as to what's played. You have people in charge, making decisions as to what to play, that are not operating under commercial constraints of consideration of how popular is this group?" (G. Gimarc, personal communication, March 8, 1995). This view was also emphasized when, in 1989, Scott Byron, editor of the College Music Journal's *New Music Report,* said that "they don't have commercial pressures. Listeners don't realize that what they hear on commercial radio has been filtered by programmers so it reaches the lowest common denominator. Yet most people are willing to take chances, to explore" (Gundersen 1989, 5D). This kind of exploration characterizes college radio. John E. Murphy, a member of the Intercollegiate Broadcasting System board of directors, then general manager of WHUS-FM at the University of Connecticut at Storrs, was quoted in 1990 as saying that "[l]arge commercial stations often don't want to give air over to extremes. They would never touch some of the stuff that is broadcast on campus stations" (Wolper 1990, 54). This, then, leads to the idea of programming "alternative" material, mainly alternative music. And, as expected, that trend has impacted commercial radio listening habits as we know them. "The shift would be the latest indicator that American's thirst for music traditionally played on college radio is growing. ... [A]n increase in fragmentation and focus on individual demographics has left traditional Top 40 high and dry" (Zimmerman 1992, 64). (See Sauls 1995.)

Of course, as has been mentioned throughout this book, the future of college radio could well be going down the commercial path. As the noncommercial band fills, college stations can look to the commercial band to place either a commercial or noncommercial signal. Realistically, though, if the station was in fact in the commercial part of the band, it would most likely elect to be a commercial, money-making entity.

> *Last I heard, there only very few (<8) successful commercial, student owned and operated radio stations in the USA. This means that the entire staff, aside from possibly a few custodians, and maybe a hired salesperson or two, are students of a university. (Airwaves ListServ, Issue Number 782, August 30, 1994)*

As of 1996, for example, many Ivy League schools have acquired commercial licenses, including Yale, Harvard, Princeton, Dartmouth, Brown, and Cornell (Harris 1996, 33). The typical commercial college station would most likely be run by hired full-time professionals, with perhaps students on board as volunteer interns. Thus, the operation of the commercial station would be taken out of the control of student programmers and staffers. This is a decision that should be considered very carefully by the college or university, weighing all options and the possible implications such a move could have on the academic program in which the station is housed.

MIRRORING COMMERCIAL RADIO

Why is it that students believe their campus radio station should sound like a commercial station? Is it because this is what they aspire to be? Is it because they believe that this will provide them a better opportunity to be hired by a commercial station, since they will then have previous experience in the same format or operation (which, actually, does make sense)? Is it because they believe more people will listen to the station? Is it because the students think they can do a better job of formatting than the local commercial station (and maybe they can, depending upon the commercial outlet)? Looking to the future, the student broadcasters can take to heart recommendations and advice given to their commercial counterparts: "For the successful radio manager of tomorrow this means developing a wider range of business skills today, particularly those involving financing and marketing for an expanding duopoly marketplace. It also means becoming computer literate in new information systems that will soon guide all station, network and rep operations" (Ditingo 1996, 37).

One of the aspects of commercial stations that students will want to undertake at the campus radio station is the remote broadcast (see Lapidus 1996). Noncommercial college stations should be very careful in carrying out such broadcasts, however, as they can tend to cross over very easily into the commercial world. One interpretation of remote broadcasts by noncommercial stations is that the station is simply inviting the listener to come out and "see" the broadcast, not necessarily enticing someone to purchase something at the site.

Technically speaking, remote broadcasts have vastly improved. Of particular note, the development of ISDN (integrated services digital network) greatly enhances the quality of the broadcast signal via the telephone. Of course, the RPU (remote pickup unit) is still a very viable transmission device. While adequate for short-range delivery (up to a 30- or 40-mile maximum depending on the antenna arrangement and power of the unit), it provides studio-quality clarity. One of the most popular RPU units is made by Marti Electronics and requires an FCC license for long-term use, and frequency coordination on a short-term basis within a given market. As always, the station should check current rules regarding FCC licensure and frequency utilization. "Telco-line" (short for telephone company line) use for remote broadcasts, of course, is not subject to such licenses or regulations.

RECOGNITION BOTH ON AND OFF CAMPUS

Although the subject of on- and off-campus audience was addressed somewhat in Chapter 8, it is revisited here to stress the importance of station recognition. As mentioned earlier, quite often the community does not understand the operation of the station, and thus recognition of the station is sometimes not properly given. Additionally, demands that are placed on the campus station are sometimes unrealistic or misleading, because the community does not know how the station functions. Broadcast requests made by outside sources are often beyond the scope of the intended purpose of the campus station (see Sauls 1996, 21). Such requests might include a sponsored (commercial) event (such as the retail remote just mentioned) or coverage of an event during holiday periods, when the station is either off the air or running with a skeleton staff.

Again, the community often confuses the campus station with commercial stations in the area. Part of this problem is that the campus radio station does not promote itself, even on campus. It is the responsibility of the college radio station to distinguish itself from other stations (usually commercial) in the area. Call letter, format, and slogan recognition are just as important in noncommercial college radio as they are in commercial radio.

WEBCASTING AND THE INTERNET

The Information Highway must be considered when discussing the future of college radio. The Internet now provides for another broadcasting outlet for radio, known as webcasting (see Beacham 1997b). Why webcasting?

> What would prompt listeners to go to all the trouble of tuning in to a webcast rather than just listening to their radios? First, webcasting allows your station to reach listeners outside of your normal coverage area. If you have programming that has any sort of universal appeal, listeners from literally around the world can and will (given the proper promotion) tune in. (Komando 1997, 44)

For those of us in college radio, this is an intriguing idea—because now we can hear what everyone else in college radio is doing. Here the concept of a global audience becomes a reality (see Haber 1996). Of course, the station will have to consider both equipment and cost factors of putting a webcast on the Internet. Also, because the college/campus radio station is not broadcasting and thus, if it is a noncommercial station, not under the noncommercial limitations as imposed by its license, couldn't it then run commercial ads in both print (on its Web site) and audio (on its webcasting)?

> In the end, probably the most important question of all is: Are any radio stations actually turning webcasting into a profit center? The answer, at least at the present time, is essentially no, although that may change as time goes on. Don't expect to generate big bucks jumping into webcasting at this point in time. *For now, look at webcasting as an investment in the future.* (Komando 1997, 44)

"Cybercast" examples can be found from AudioNet, a Dallas-based company providing numerous stations and stored audio files. Their URL is *http://www.audionet.com*. A listing of public/noncommercial Web radio stations, with links to listen to their broadcasts, can be found at the BRS Web Radio site *http://wwww.web-radio.com/fr_public.html*. Additionally, the "*NAB Websource* is an interactive guide designed to keep stations up to date about Internet developments that will have an effect within the broadcast industry." As of 1998 the site was "accessible to NAB members only, but will be made available for prospective members and interested parties for a one-month trial period" (NAB debuts Websource, 1998, 73).

Also, as mentioned in Chapter 7, McClung's 1998 doctoral study centered on the uses of college radio Web sites (personal communication, October 5, 1998). Finally, as to be expected, legal aspects of Internet broadcasting are only beginning to be investigated (Slater 1997). Future judgments and rulings could drastically impact the use of the Internet in regard to broadcasting applications. Items yet to be determined at the time of this writing include such topics as music clearance and licensing for use over the Internet, along with links to existing sites by radio stations. At the present time the FCC does not regulate "broadcast" stations established solely for distribution over the Internet or on the World Wide Web. (Information pertaining to Internet broadcasts or "webcasting" may be obtained via the Federal Communications Commission Audio Services Division Web site at *http://www.fcc.gov/mmb/asd/main/other.html.*)

PRACTICAL APPLICATIONS

Creating a WWW Site

Web-site advertising is a revenue-generating source for college radio stations. Here the reader is directed to Appendix C for an explanation on creating such a Web site. Entitled *Creating a World Wide Web Site for a College Radio Station: Considerations, Concerns, and Strategies,* a discussion is presented by Craig A. Stark (1997) from Sam Houston State University. Such points as equipment, monthly costs, maintenance, and so on, need to be explored when considering any move to a new technology. (See also Skip Pizzi's "An On-Line Radio Primer" in the February 1998 issue of *BE Radio.*)

College Radio and Digital Audio Technology

As digital audio evolves, it is only natural that college radio stations undertake the same industry standards. Here, then, I address the following issues: (1) the need for and utilization of digital audio in the college radio station; (2) the selection of digital audio systems; and (3) applications to foster the teaching of digital audio (see Sauls 1997). Additionally, a checklist is provided for use when considering the acquisition of digital audio systems.

Digital Audio in the College Radio Station

If you are the manager of a college radio station, it is inevitable that you are facing the dilemma of which digital audio system to acquire. This matter is no longer one of "Are we going to do it?" but rather, "We must do it to keep our students and station up-to-date with the industry technology." "Although they are but a portion of station operations," argue three professionals in the field, "digital production tools can help you shape the face of your station image, and productions. They can also take you to the next level of sound, quality, and creativity in your media" (Barker, Pierce, and Albright 1996, 18).

From the outset, it is important to understand what the industry is talking about when we use the term "digital audio." I propose that two distinct, yet overlapping, areas are encompassed in digital audio at the radio station: programming and production. Concerning programming, the digital system houses the on-air product (music, spots, jingles and sounders, news and sports actualities, etc.) and thus replaces many of the analog devices (in particular cart machines, which are moving to minidiscs, turntables, and CD players as material is being loaded to hard drives). The amount of product digitally stored depends on the memory capability of the computerized system being used. Of course, one must be aware that such digital programming will most likely play very heavily in the future within the context of Internet radio (see Beacham 1997a), cable radio, DAB (digital audio broadcasting), and DTH (direct-to-home satellite radio).

Concerning the use of digital audio in the actual broadcast/ transmission application, Barker's 1996 article, "Basic Anatomy of a Digital Broadcast," has the following recommendation to make: "[D]igital technology revolution has more to offer your station than just production and delivery. It can provide the basis for the cleanest, best sounding radio broadcast available anywhere. It impacts how the radio signal is processed and generated, and how well your station is heard by your listeners" (15). This article provides a good overview of digital technology and the broadcast chain. "Exploring the possibilities of going digital is a very worthwhile endeavor for an institution. Analog broadcast components were not built to last forever and wear out over time" (18). As mentioned in Chapter 3, automation, satellite-fed programming, and unattended operation now allow for true "people-free" radio stations.

Another important consideration in digital audio broadcasting is the factor of the noncommercial band in the current waveform/analog transmissions of today. When discussions ensued concerning the creation of digital audio broadcasting in the early 1990s, the National Association of College Broadcasters "emphasized the need to continue the policy of reserving noncommercial channels under DAB. Compared to their commercial counterparts," wrote NACB, "college stations generally will have more difficulty purchasing the required new technology. A reserved band will help insure that they will eventually be able to participate in the digital age once the requisite finances are obtained (National Association of College Broadcasters 1995, 69). Information pertaining to digital audio radio services by satellite may be obtained through the Federal Communications Commission Audio Services Division Web site at *http://www.fcc.gov/mmb/asd/main/other.html*.

As for production, Skip Pizzi writes in the January/February 1997 issue of *BE Radio* (of which he is the editor) that "the traditional use of a series of discrete devices for record/playback, mixing and processing of audio is being replaced by a more monolithic single device—the digital audio workstation (DAW)" (24). He explains: "It is fairly well-known that DAWs are available in three basic types, defined by the computer upon which they are based: the Apple Macintosh, the IBM PC or a nonstandard ('proprietary' or 'dedicated') computer. Among the many variations between DAW systems, this is the most basic distinction" (26).

All concerned would probably agree that when considering digital systems for either programming and/or production, they would prefer those that have the following attributes: intuitiveness, power, ease of use, ease of configuration, little need of maintenance, reliability, and a minimal learning curve. In reality, for many of us in the academic environment, most equipment considerations are based on economics. So if we can acquire systems that will run on already existing computer equipment at the school, we are halfway there. Of course, use of LAN systems on campus (local area networks) can provide the tie between production and programming, if needed.

In regard to the acquisition of "proprietary," systems, one should note with caution that such systems can prove hard to justify, since they are stand-alone units and do not provide for use with existing systems and cross-utilization by others. But proprietary systems are at times specifically warranted, since they are totally dedicated to your use and mission. Additionally, proprietary systems do not require that the system "fit in" with existing hardware, which can limit or alter your intended functions.

It is also important to ensure the compatibility of any digital system acquired with the computer automation existing in the station. "By 1999, computer automation is expected to be used in some form by virtually every radio station in markets big and small." (Rusk 1996; See also Jones 1996 concerning traffic and billing systems manufacturers.)

Selecting a Digital Audio System

Once the decision has been made to move into, or even upgrade an existing, digital audio system, then the matter becomes one of selection. A specific consideration for those of us in the academic world is that whatever piece of equipment we choose, we are aware that it will be around for a long time. It is not uncommon to see pieces of analog equipment at a campus radio station that are close to 20 years old. Reality tells us that whatever system we choose will define what we will be working with and teaching on for years to come.

The fact is that there are a lot of systems to choose from. In Chapter 1, I wrote that "as with computers, what appears current today was probably outdated by a new development yesterday." I recommend that you educate yourself on not only specific systems, but digital technology overall. (Some useful references: the USA Digital Radio home page at *http://www.usadr.com.*, and the *Radio and Production* magazine at *http://www.rapmag.com.*)

When choosing a system, it is imperative that all possible uses be considered. Because this system may be housed in an academic environment (one in which teaching also takes place), be sure to review the use of the digital audio system in conjunction with video and/or film sound applications. Thus, does your digital audio system need to have SMPTE time code ability? Will your system have a 24 fps (frames per second) need? What peripheral equipment (analog and digital) will be used for loading and storing audio with the digital audio system, and how will that equipment be connected to the system?

As for cost, don't cheat yourself! Short-changing your station and/or academic department can have a major impact on your station and program development. Up-front costs may be high, particularly if you need to purchase a lot of hardware. But going the more competitive, lower-priced route may not be the best decision initially. You may have to work on justifying a more high-end system, but it will be worth the effort.

As with any studio component (equipment, furniture, or computer-based systems), it is important to address both the practical use and aesthetic appeal prior to purchasing and installation. First and foremost for the staff, always study ergonomic factors (the human element). Concerning an installation, Richard Schrag comments on the importance of studio and facility design: "To define a space plan with adjacencies and traffic flow suited to the complex day-to-day operations of a radio station … [p]articular attention was paid to sound isolation, the acoustical environment in the studios, HVAC noise and vibration control, comprehensive wire management systems, and the accessibility and storage of materials and equipment. Every decision was evaluated in the context of the facility's long-term flexibility"(1997, 88). In sum, he states that "[g]ood planning and solid design yield a foundation that will stand the test of time" (90).

A final thought concerning the selection of a digital audio system has to do with the basics of investing in computers. Kevin McNamara has addressed common mistakes undertaken with acquiring PC-based technology. These can easily be adopted when considering a digital audio system. He writes that:

> Profitable investment in computer equipment follows a different set of rules, and like many other industries, broadcasters are learning this the hard way. Consider the following "seven deadly sins" of computer asset management, and see if at least some don't seem painfully familiar:
>
> 1. Lack of a strategic plan.
> 2. Placing the engineering department in charge of all computer systems.
> 3. Investing in the wrong equipment and trading computer equipment.
> 4. Employees specifying equipment.
> 5. Lack of a software standard.
> 6. Improper training.
> 7. No security plan. (1997, 12, 14)

(For use in the selection of a digital audio system, a checklist is provided along with examples of software and hardware later in this chapter.)

The Teaching of Digital Audio

As we move into the digital domain, we as teachers of audio/radio production are ironically finding ourselves returning to the world of analog concepts. Basically, we are discovering that though students can use a computer, that doesn't make them production literate. The concepts of audio (sound) and radio production (mixing, layering, transitioning, etc.) are best learned—for whatever reason—in the analog world. Some of us believe that because the student is sitting in front of an audio console and working with physical tape and equipment, a more conducive atmosphere to production exists. (That is why many digital audio systems tout the ideal that you are still sitting in front of a mixer, and so already somewhat familiar with the atmosphere and not scared by the technology.)

Reality also shows us that not all radio stations are moving so quickly into the digital world. Let's face the fact that the last place a lot of station managers and owners want to put money is into equipment. So we just can't completely turn to teaching digital when the student will probably be confronted with analog for some time to come in the commercial (real) world. But as more stations (both commercial and noncommercial/public) move to digital use, we must also be teaching in that arena.

What do we teach concerning digital audio? My own experience (endorsed by professional studio producers) is that the understanding of digital concepts and theory is more important than learning a specific system. Here the ideals of track and memory management are continually reinforced because of the domain in which digital audio operates. Theoretically, if the student can learn the underlying concepts of digital audio, those concepts can be applied across all systems. In short, if a student just learns "a system," that student will be able to work only on that specific model. But if the student can understand the theories of digital audio, these can be applied to all systems. And what are the chances that the system one learns on at school will be the same system at one's first internship or first job?

Of course, the practical understanding of digital audio must be taught on a system. Additionally, the radio station will be operating on a given system. As a teacher, either in the classroom or at the college radio station, remember that while you are teaching a computerized system, you also need to teach the basics of audio production.

Getting Closure in the Digital Domain

Finally, we must consider the aspect of completing a project or a production and walking away from it. We are finding more and more that in the use of digital technology (including video and film applications) producers are finding it more difficult to gain closure on productions because of the ability to continually "fine-tune" various aspects. The ideal of "correction to perfection" because of the power of the technology begins to reach the point of absurdity. The student as producer must be able to find a reasonable point of contentment: The concept of reworking over and over again to improve the work needs to be controlled. Otherwise, one is never finished or satisfied with one's work.

The digital process allows you to work with sound more precisely, and grants you more creative freedom and efficiency in producing spots. It's important to remember, however, that it's not only the computer that directly makes your production better. It will always be the talent, not the tools, that gets the job done. (Barker, Pierce, and Albright 1996, 14)

Learn how to utilize gracefully these powerful tools in the station!

Checklist

Use:
Programming and/or Production

Attributes:
Intuitiveness, Power, Ease of Use, Ease of Configuration, Little Maintenance Required, Reliability, Minimal Learning Curve

Cost:
(May be predetermined by station budget or capital allocation.)

System Needs and Compatibility:
Compatibility with existing computer equipment?
Campus LAN system compatibility?
Proprietary system?
Video and/or Film sound applications?
SMPTE time code needed?
24 fps (frames per second) needed?
Peripheral equipment (analog and digital) compatibility?

The Orban DSE 7000 Workstation Worksheet contains the following concerning selection:

Ease of Use:
"Look and Feel"
Editing Ease
Training and Installation

Speed and Productivity:
Editing Speed
System Speed
Return on Investment

Software and Hardware Examples (as of 1998)

Macintosh Platform:
Macromedia's sound package SoundEdit 16
DigiTrax 1.2 multitrack recording

Macromedia's Deck II 2.5
Digidesign's ProTools DAE Powermix
Digidesign's Audiomedia II and III
Digidesign's ProTools III and 4.0 systems

Windows/PC Platform:
Innovative Quality Software's SAW and SAW Plus
Digital Audio Labs' FastEddie and EdDitorPlus

Source: Barker, Pierce, and Albright 1996

QuickTime 3.0 Audio Tools (Macintosh Based)
Source: Florio 1998, 57

Others:
Multitrack Editing: Digidesign Session
SAW, Software Audio Workstation
Cool Edit and Cool Edit Pro

Proprietary Systems Noted by the Author:
Roland DM-80 and DM-800
Harris-Allied DSE 7000 (formerly Orban)

REFERENCES

Barker, J. S.(1996). Basic anatomy of a digital broadcast. *College Broadcaster* 10(2): 15, 18.
Barker, J., L. Pierce, and B. Albright. (1996). Tools in transit: New technologies for innovative production. *College Broadcaster* 10(2): 14, 16-18.
BE Radio. (1998). November/December 4(10).
Beacham, F. (1997a). Internet broadcasting booms. *Radio World* 21(1): 20-21.
———. (1997b). Yes, Virginia, it sure is radio! *Radio World* 21(14): 21, 25.
Ditingo, V. M. (1996). Management journal: Learning curve. *Tuned In* 3(2): 37.
Florio, C. (1998). Rapid review: QuickTime 3.0 audio tools. *Pro Audio Review* 4(8): 57.
Gundersen, E. (1989). College radio explores rock's flip side. *USA Today,* February 27.
Haber, A. (1996). It's a world wide web for radio. *Tuned In* 3(5): 25-30.
Harris, L. (1996). Ivy League broadcaster bills big bucks in JSA. *Radio World* 20(12): 33, 37.
Jones, S. (1996). Automate your way around traffic jams. *Radio World* 20(4): 38, 42.
Komando, K. (1997). My wonderful webcasting adventure. *Tuned In* 4(9): 44.
Lapidus, M. (1996). Do you take your DJ out in public? *Radio World* 20(14): 44.
McNamara, K. (1997). Preserving computer investments. *BE Radio* 2(1): 12-14.
NAB debuts Websource. (1998). *BE Radio* 4(8): 73.

National Association of College Broadcasters. (1995). *1995 NACB station handbook.* Providence, R.I.: National Association of College Broadcasters.

Pizzi, S. (1997). Creating content. *BE Radio* 2(1): 22-30.

———. (1998). An on-line radio primer. *BE Radio* 4(2): 28-32, 38-42, 66.

Rusk, B. (1996). Forecast predicts new uses for radio. *Radio World* 20(5): 3, 6.

Sauls, S. J. (1995). College radio. Paper presented at the 1995 Popular Culture Association/American Culture Association National Conference, Philadelphia, April 14, 1995. ERIC, ED 385 885.

———. (1996). College radio: Points of contention and harmony from the management perspective. *Feedback* 37(2): 20-22.

———. (1997). Preparing students to enter the digital age: College radio and digital audio technology. Invited paper presented at the paper/poster session of the Student Media Advisor's division at the 1997 Broadcast Education Association (BEA) annual convention, Las Vegas, Nev., April 4, 1997. ERIC, ED 410 627.

Schrag, R. (1997). Designing for consolidation. *BE Radio* 2(2): 88, 90.

Slater, E. S. (1997). Broadcast on the Internet: Legal issues for traditional and Internet-only broadcasters. *Media Law and Policy Bulletin* 6(1): 25-42.

Stark, C. A. (1997). Creating a World Wide Web site for a college radio station: Considerations, concerns, and strategies. Paper presented at the Texas Association of Broadcast Educators Conference, March 1, 1997, Dallas.

Wolper, A. (1990). Indecency suit chills campus stations. *Washington Journalism Review* 12(9): 54.

Zimmerman, K. (1992). Alternative pops pop's balloon. *Variety,* March 9, pp. 64, 66.

10

Final Practical Applications

This chapter is an attempt to bring together numerous ideals within the context of programming and managing a college radio station. Ultimately, as a faculty advisor, station manager, or program director, you will find that these final points highlight what can be done to foster "harmony" within the staff of the operating college radio station (see Sauls 1996). And it is in this spirit—the positive—that these points are presented.

COMMUNICATION

Many of the "problem areas" addressed in this book can be curtailed somewhat through ample and appropriate communication. Confusion over the needs, wants, and desires of the college radio station staff is often the result of inadequate or misleading information. The more enlightened the station student staff and outsiders (both on and off campus) are as to the operation of the station, the more they will be able to comprehend the manager/advisor's stated purposes and intended actions. By communicating information appropriately, the station manager/advisor might find him- or herself conveying more and justifying less. (Oh, the time we spend over problems that could have been neutralized with adequate communication.)

One recommendation in preventing a crisis is to "[m]ake a list of everything that could become a crisis at your station. Consider the possible consequences, and estimate the cost of prevention. Develop a step-by-step strategy for how you would deal with potential problems" (Jones 1996, 46). This will be a philosophy of anticipating situations, and thus of being ready to handle them.

INVOLVEMENT AT THE COLLEGE RADIO STATION

Get your staff involved in all aspects of the station ... office work, promotion, production, formatics, news, public affairs. Work with them on the content of their breaks as well as their musical content. Once they have a well-rounded understanding of radio you may have fewer problems. If you don't get them on board philosophically, no amount of money will bring them around. (NACB ListServ, September 26, 1997)

GET INVOLVED

I really enjoy reading The Radio World Magazine, especially the letters from people interested in gaining experience in broadcasting. I cannot overemphasize the importance of students getting involved in activities at the college radio station.

At Rowan College in New Jersey we have more than 150 students involved in different areas at WGLS-FM. All students must go through a five-week training program before becoming a member. This program covers basic FCC rules and regulations, station policies and procedures, and includes an introduction to production techniques. The station is structured with students as department heads, giving them a basic introduction to managing people. To gain additional skills we strongly urge the student to participate in an internship at a commercial station in the area. I also recommend active participation in the National Association of College Broadcasters (NACB) as a way to learn more about all the aspects of station operations.

Frank J. Hogan
General Manager
WGLS-FM
Glassboro, N.J.

(Hogan 1995, 9. Used with permission.)

UNDERSTANDING STUDENT LIMITATIONS

The manager/faculty advisor must have an understanding of student limitations. These students are individuals who may never have been in a station until they walked into the campus radio station. Quoting Dr. Roosevelt Wright, Jr., a professor at Syracuse's Newhouse School of Public Communications in the radio/television/film sequence: "Students have to be able to make mistakes. ... The college or university provides an ample opportunity for a person to make those mistakes. Having been a broadcaster myself—and a person who's owned broadcast properties and managed them—usually we don't have time to train someone in the industry" (Chichester 1995, 62).

It's not that the manager/advisor is encouraging a lesser quality of work by the students themselves but, rather, is tolerant of expected errors and blunders. Most important, the supervisor must accept the fact that the same mistakes will be made over and over again as staffs rotate (see "Staffing" in Chapter 6). What seems obvious to those of us with years of experience in the radio industry is something completely new for the student novice.

STRIVE TO BE POSITIVE

Always strive to be positive, even in the worst of times. Students tend to emulate those who direct them. Bad attitudes will only serve to create negative feelings. No matter what happens, always try to communicate the positive. Even if the radio station catches fire! (Yes, this actually happened at the college radio station I managed. We communicated to our student staff that this is just something that happens and that we would have to continue to operate no matter what. Our engineer had us operating the day after the fire in a makeshift studio. While we were out of our building for just under a year, we did get our facilities renovated, and all new equipment. And, in all honesty, the disaster provided a great public relations tool for the station!)

CREATE A PROFESSIONAL ATMOSPHERE

Do your best to create a professional atmosphere. If the student staff are placed in a professional operation, their performance will be of a professional nature. At the same time, though, be sure to allow enjoyment and experimentation. Of course, this carries over into the educational/learning aspect within the station itself (as addressed later in this chapter). (See "Good Morning, Class" in the February 1998 issue of *Tuned In,* with the cover title "Reading, Writing and Arithmetic-AM-FM: Educating Tomorrow's Radio Broadcasters.")

GIVE RESPONSIBILITY

Give responsibility in order to further nurture the creative activity. Remember that students are on campus (and at the campus station) to learn. But they must be given the opportunity to do their job at the station. It is the station manager's job to manage and advise, while letting the student staff run the station.

Just because the staff may consist of volunteers doesn't mean they don't want responsible positions. As Peter F. Drucker, one of today's management "gurus," wrote in a 1998 issue of *Forbes* magazine: "What motivates—especially knowledge workers—is what motivates volunteers. Volunteers, we know, have to get more satisfaction from their work than paid employees precisely because they do not get a paycheck. They need, above all, challenge. They need to know the organization's mission and to believe in it. They need continuous training. They need to see results" (166).

LEARNING

As is expressed in the following letter, one never stops learning, even in the professional world.

LEARNING
AN ONGOING PROCESS

Contacts with college students have been discouraging the past few years, but the "Letters" section in the September issue of The Radio World Magazine renewed my hopes. Your replies to both of the students' questions were excellent! May I add a couple or more thoughts?

Where does a young person get experience? You never stop. After forty-six years in radio, it's still a learning game for me, and that's what makes it fascinating. Changing rules and regulations, new technology, dealing with a personality you've never encountered … it never ceases.

Learn to evaluate advice. When we old codgers say it can't be done that way, it just means that we have not been able to do it. Try out new ideas. Don't be afraid of failure. Look upon efforts that are not successful as the best way to learn.

Most importantly, learn cost accounting. Most mistakes I've made have been reacting emotionally rather than logically. Do you really need that little box that will make you sound better?

Jim Farr
General Manager
KKUB(AM)
Brownfield, Texas

(Farr 1995, 6. Used with permission.)

ESTABLISH YOURSELF AS MANAGER,
WITH CONSIDERATION

Whether you're the station faculty advisor, full-time staff program director, or station manager of the college radio station, it is important to establish yourself as the director. This is not an authoritarian or dictator role (which, by the way, can sometimes be found in the college radio station), but one of leadership. Students need a leader whom they can trust and turn to—one who knows the operation and is responsible for its continuation. Make clear to the students that their actions will determine the future of the station. This helps them to believe in themselves and in the station.

As mentioned earlier in this chapter, be sure to understand (and never forget) that you are working with students, who do have feelings (in addition to limitations). It has been my experience in managing college radio stations (and even more important, as a faculty member) that students can be very fragile. They will take to heart your off-the-cuff comments. It is important that the station leaders (manager, program director, and/or faculty advisor) be very careful in their general comments and express to their staff that they should always feel free to approach the manager with any questions or concerns they may have. Here the station leaders are showing the same respect and consideration that they would expect to receive from their own supervisor, subordinates, and counterparts.

PRACTICAL APPLICATIONS

Sensible Operations

As almost an epitaph: "Common sense" should play a role in the everyday operation of the college radio station. In the style of commentary, I cannot overly emphasize the idea that logical and sensible decisions will assist greatly in the carrying out of station matters.

Time spent on making choices in staffing, purchasing, program acquisition, and so on, will always pay off. This is also where "gut feelings" come into play. Most likely those making decisions will have some type of inherent ability to see projects and tasks to completion.

The FCC, college and university administrations, city officials, departmental chairs, academic colleagues, and community citizens, all believe that the station director (be it a faculty advisor or staff manager) has the ability and knowledge to run the college radio station. People (listeners included) respect the ongoing operation of the station and admire staff perseverance day after day, even if it is never acknowledged.

Finally, within the discussion pertaining to student training, as outlined in Chapter 6, a point needs to be made that today more and more students are training fellow students. This is only natural in the close-working environment within a college radio station, coupled with the limited number of advisors or staff to "show the way." Of course, the more advanced students (seasoned pros) will assist in helping to guide fellow beginners. This, in turn, provides the mature student with the ability to take responsibility, and possibly consider station management or even teaching as a future career choice.

REFERENCES

Chichester, P. (1995). Education vs. experience. *Radio World Magazine* 2(8): 58-64.

Drucker, P. F. (1998). Management's new paradigms. *Forbes,* October 5, pp. 152-76.

Farr, J. (1995). Learning an ongoing process. *Radio World Magazine* 2(11): 6.

Good morning class: Educating tomorrow's radio broadcasters. (1998). *Tuned In* 5(2): 8-15.

Hogan, F. J. (1995). Get involved. *Radio World Magazine* 2(11): 9.

Jones, S. (1996). What to do when crises strike. *Radio World* 20(14): 46.

Sauls, S. J. (1996). College radio: Points of contention and harmony from the management perspective. *Feedback* 37(2): 20-22.

Program Listings Available

Randall Davidson, Chief Announcer at Wisconsin Public Radio, has compiled an exhaustive list of programs available via numerous sources. As addressed in Chapter 3, this list showcases the vast selection of programs available, particularly those open to noncommercial stations by means including traditional (tape, compact disc, etc.) and satellite distribution. Mr. Davidson's broadcasting career began at WRST-FM at the University of Wisconsin-Oshkosh, and he later worked at WHBY-AM in Appleton, Wisconsin.

PROGRAMMING SOURCES

Alphabetical by program title.
(+) = also available on NPR satellite.
Please send updates/corrections to Randall Davidson at:
davidson@vilas.uwex.edu.
Please let producers know where you learned about their program.

A-INFOS RADIO PROJECT
Variable-length, spoken word programs (interview segments, news modules, etc.) from various sources available for download from Web site:
[*http://www.radio4all.net*]

AMERICAN CANCER SOCIETY RADIO NEWS SERVICE
Four monthly radio voice reports on cancer research.
*FREE by telephone: 800-276-6397
American Cancer Society: 212-382-2669

AMERICAN MEDICAL ASSOCIATION RADIO NEWS
Daily radio voice report on health and medical issues.
*FREE by telephone: 800-448-9384
AMA Radio News: 312-464-4449

Dated: September 10, 1998; Updated October 21, 1998. Used with permission.

AMOCO/CHICAGO SYMPHONY ORCHESTRA
120-minute, weekly classical music concert performances (+)
On 10.5″ open reels or DAT—tape return
Contact producer for fees
WFMT Fine Arts Network: 773-279-2110

THE BAPTIST HOUR
28-minute, weekly religious program
*FREE on CD
Media Technology Group/Southern Baptist Convention: 800-266-1837

BEALE STREET CARAVAN
30-minute, weekly program of live blues performances recorded at festivals
  and clubs around the country. The program also includes a celebrity
  host talking about their area of expertise.
*FREE on cassette. 40 new programs per year and 12 repeats for full-year
  service.
Stations must demonstrate and verify consistent carriage.
The Blues Foundation: 901-527-2583

THE BEST OF OUR KNOWLEDGE
25-minute, weekly program featuring conversations with college faculty
  members about academic topics (produced by WAMC-Albany). (+)
$5.00 per program on cassette (plus ordering fee)
Longhorn Radio Network: 800-457-6576

THE BIG BACKYARD
30-minute, hosted music program featuring new musical artists from
  Australia. The program features music from a variety of styles and
  genres.
28 programs per year. On CD.
The Big Backyard GPO 697, Sydney 1043, Australia: 61-2-9360-4547
  bby@next.com.au

BLUES BEFORE SUNRISE
5-hour, weekly program of blues music
*FREE on DAT—tape return (+/archive tapes of satellite program)
Stations must agree to broadcast program as a five-hour block. Not available
  in markets having stations carrying program via satellite.
Steve Cushing: 708-771-2135

BLUES FROM THE RED ROOSTER
59-minute, weekly program of blues music
$13.00 per program on cassette (plus ordering fee)
Limited market availability—contact distributor for details
Longhorn Radio Network: 800-457-6576

BODYTALK

59- or 29-minute, weekly program of health and medicine, featuring news features, topical discussions, and prearranged calls from listeners (listeners can call-in to an answering machine and are called back during taping, so stations can run the program at any time) (Produced by WOSU-Columbus, Ohio) (+)

*FREE on cassette or DAT

Stations must sign a broadcast agreement and agree to air funding credits, and notify the producer if they stop airing the program.

Susan Gorman/WOSU: 614-292-9678

THE BOOK SHOW

29-minute, weekly literary program featuring author interviews and discussion about books from best-sellers to the classics. (+)

Produced by WAMC-Albany

*FREE on cassette. DAT available. Contact producer for details.

National Productions: 1-800-323-9262, ext. 165

BOSTON ROMANTICS

Four programs highlighting the recordings of late-19th century American composers from the Boston area.

$195.00 for series on DAT or analog tape (no return, may be rebroadcast)

Radio Features Corporation: 301-718-2800

BY THE WAY

90-second daily module of inspirational "thought for the day"

*FREE on cassette or CD

The Jubilee Network: 800-325-6333

CAMBRIDGE FORUM

29-minute, weekly program of issues/policy/history discussion (+)

$5.00 per program on cassette (plus ordering fee)

Longhorn Radio Network: 800-457-6576

CAPITOL CONNECTION

29-minute, weekly interview program featuring New York State politicians. Although of primary interest to New York residents, it is available to all noncommercial stations. (+)

Produced by WAMC-Albany

*FREE on cassette. DAT available. Contact producer for details.

National Productions: 800-323-9262, ext. 165

CELTIC CONNECTIONS

59-minute, weekly hosted program of Celtic music recordings

Market exclusivity. (Produced by WSIU-Edwardsville, Illinois) (+)

*FREE on DAT. Contact producer for shipping fees.

WCLV: 1-800-491-8863, Attention: Tara Renk

CLEVELAND CITY CLUB FORUM

59-minute, weekly public affairs program featuring the speakers at the Cleveland City Club. Program consists of speaker's address for ~30 minutes followed by audience questions. (+)
*FREE on DAT. Contact producer for shipping fees.
WCLV: 800-491-8863, Attention: Tara Renk

CLEVELAND CLINIC RADIO NEWS SERVICE

Weekly voice report and actualities on health issues
*FREE via phone: 800-428-0050

CLEVELAND ORCHESTRA

~90 minute weekly classical music concert performances (26 weeks/Jan-Jun) (+)
*FREE on DAT. Contact producer for shipping fees.
WCLV: 800-491-8863, Attention: Tara Renk

COLLECTOR'S CORNER WITH HENRY FOGEL

20-minute, weekly program of classical recordings, chosen by Chicago Symphony president Henry Fogel. (+)
On 10.5" open reels or DAT—tape return. Contact producer for fees.
WFMT Fine Arts Network: 773-279-2110

CONCERT HOUR

57-minute, weekly program of classical music performance excerpts. 26 weeks per year. (Produced by Radio Deutsche Welle) (+)
On 10.5" open reels or DAT—tape return. Contact producer for fees.
WFMT Fine Arts Network: 773-279-2110

COUNTERSPIN

28-minute, weekly program of media criticism and media activism on cassette (+)/quarterly fee/sliding scale
Contact producer for fees
Fairness and Accuracy in Reporting: 212-633-6700

COUNTRY CROSSROADS

28-minute, weekly program of country music and interviews with country music artists, along with inspirational commentary from guests and hosts
*FREE on CD
Media Technology Group/Southern Baptist Convention: 800-266-1837

DETROIT SYMPHONY ORCHESTRA

120-minute, weekly classical music concert performances (26 weeks) (+)
*FREE on DAT. Contact producer for shipping fees.
WCLV: 800-491-8863, Attention: Tara Renk

DEUTSCHE WELLE FESTIVAL CONCERTS
120-minute, weekly classical music concert performances (13 weeks per
year) On 10.5" open reels or DAT—tape return (+)
(Produced by Radio Deutsche Welle) Contact producer for fees.
WFMT Fine Arts Network: 773-279-2110

EARTH AND SKY
90-second, daily program on scientific news of the natural world
*FREE on CD
Earth and Sky: 512-477-4441

EARTHWATCH RADIO
2-minute, weekday modules on the environment, ecology, water resources
*FREE on cassette/no excepting/must air funding credit
Institute for Environmental Studies, University of Wisconsin: 608-263-3063

THE ENVIRONMENT SHOW
25-minute, weekly newsmagazine on the environment and ecology
(Produced by WAMC-Albany) (+)
$5.00 per program on cassette (plus ordering fee)
Longhorn Radio Network: 800-457-6576

EUROPEAN CENTURIES
59-minute, weekly classical music performances from European orchestras
(produced in association with the European Broadcasting Union) (+)
*FREE on DAT/contact producer for shipping fees
WCLV: 800-491-8863, Attention: Tara Renk

EUROQUEST
28-minute, weekly magazine program with news and features from Europe
*FREE on CD (CD includes RADIO NETHERLANDS DOCUMENTARY)
Excerpting of modules allowed with credit to Radio Netherlands Radio
Netherlands: 800-797-1670

EVANGELICAL LUTHERAN CHURCH IN AMERICA NEWSLINE
Weekly radio feeds (produced spots)
*FREE by telephone: 800-446-3975
ELCA: 312-380-2958

FAMILY TALK WITH SYLVIA RIMM
59- (or 30-) minute, weekly call-in program on parenting issues (listeners
can call in to an answering machine and are called back during taping,
so stations can run the program at any time) (+)
*FREE on DAT. Contact producer for shipping fees.
WCLV: 800-491-8863, Attention: Tara Renk

FIFTY-ONE PERCENT

25-minute, weekly magazine program on women's issues and news

$5.00 per program, on cassette (plus ordering fee) (+) excerpting of modules allowed with credit (Produced by WAMC-Albany)

Longhorn Radio Network: 800-457-6576

THE FIRST ART

59-minute, weekly program showcasing the heritage of the world's vocal music traditions. Stations must sign and return broadcast agreement (Produced by KUSC-Los Angeles) (+)

*FREE on DAT. Contact producer for shipping fees.

WCLV: 800-491-8863, Attention: Tara Renk

FIRST HEARING

54-minute, weekly panel discussion of new classical recordings (+)

$25.00 per program, on tape, tape return basis

Radio Features Corporation: 301-718-2800

FM ODYSSEY

3-hour, weekly program featuring music of acoustic singer, songwriters with interview segments. Music spans eclectic rock, blues and folk (+)

*FREE on DAT or cassette

Some material may be date-specific. Contact producer for other restrictions.

Fred Migliore: 800-341-1480

FORUM

29-minute, weekly program of current affairs, cultural topics, science (Produced by University of Texas-Austin) (+)

$4.50 per program on cassette (plus ordering fee)

Longhorn Radio Network: 800-457-6576

GETTING A LIFE

29-minute, weekly program of conversations on everyday entrepreneurship and how to find the right livelihood (13 programs in series) (Produced by WOSU-Columbus, Ohio) (+)

*FREE on cassette or DAT

Stations must sign broadcast agreement and agree to air funding credits, and notify producer if they stop airing the program.

Susan Gorman/WOSU: 614-292-9678

GRACE NOTES

2.5-minute, weekday modules of stories drawn from the lives of classical composers and performers

*FREE on CD

Public Broadcasting Radio Service: 781-595-3747

THE HEALTH SHOW
25-minute, weekly newsmagazine on health, medicine, fitness
$5.00 per program on cassette (plus ordering fee) (+)
Excerpting of modules allowed with credit (produced by WAMC-Albany)
Longhorn Radio Network: 800-457-6576

HEALTHBEAT RADIO NETWORK
60-second spots on health from the National Heart Lung and Blood Institute
of the National Institutes of Health
39 programs per quarter
*FREE on CD (with host intros)
Jameson Broadcast: 202-338-4800

HIGH PLAINS NEWS SERVICE
14-minute, weekly magazine program on rural issues in the western U.S.
On cassette (+), sliding fee schedule. Contact producer for information.
Excerpting of modules allowed with credit.
High Plains News Service: 800-729-3540

I WRITE THE SONGS
29-minute, weekly seminar program on songwriting with instruction, critiques and songwriting examples
$8.00 per program on cassette (plus ordering fee)
Longhorn Radio Network: 800-457-6576

IN BLACK AMERICA
29-minute, weekly program highlighting black life and culture in American society (Produced by the University of Texas at Austin)
$4.50 per program on cassette (plus ordering fee)
Longhorn Radio Network: 800-457-6576

INFINITE MIND
60-minute, weekly program exploring the human mind and psychological issues
$7.50 per program (plus ordering fee), on cassette
Longhorn Radio Network: 800-457-6576

INNOCENTS ABROAD
29-minute, weekly program of travel-related interviews and travelogue sound portraits
*FREE on cassette
Innocents Abroad Radio Productions: 707-974-7821

INSIGHT

2- to 3-minute, weekday commentaries from Matthew Rothschild, editor of Progressive Magazine

*FREE on cassette

The Progressive: 608-257-4626

THE JAZZ DECADES

59-minute, weekly program of jazz music and jazz history hosted by jazz musician Ray Smith (Produced by WGBH-Boston) (+)

$7.50 per program on cassette (plus ordering fee)

Not available to commercial stations

Longhorn Radio Network: 800-457-6576

JOY

54-minute weekly program of traditional church music

*FREE on CD (program contains commercials as separate tracks on CD: noncommercial stations can skip these tracks)

The Jubilee Network: 800-325-6333

KITKAT ACOUSTIC BREAK

30-minute, weekly program with musician interviews and studio recordings in an acoustic setting (20 programs per year: 10 in fall, 10 in spring)

*FREE on CD

Jaime Levit/College SoundTrack: 212-921-2100

THE KOSTRABA CONUNDRUM

57- to 59-minute, weekly hosted program of 20th-century music

*FREE on DAT (Produced by WBAA-West Lafayette, Indiana)

Greg Kostraba: 765-494-5920

LATINO USA

29-minute, weekly magazine of news and culture with a Latino focus (+)

*FREE on cassette, magazine and modular formats on same tape, excerpting permitted

Longhorn Radio Network: 512-471-1817

THE LAW SHOW

25-minute weekly program on the courts and legal issues

$5.00 per program on cassette (plus ordering fee) (Produced by WAMC-Albany) (+)

Longhorn Radio Network: 800-457-6576

LEGISLATIVE GAZETTE

29-minute, weekly newsmagazine devoted to covering New York State
Legislature and New York State government and politics. Although of
primary interest to New York residents, it is available to all noncommercial stations. Produced by WAMC-Albany (+)

*FREE on cassette. DAT available. Contact producer for details.

National Productions: 800-323-9262, ext. 165

LIFETIMES: THE TEXAS EXPERIENCE

90-second, weekday program of history and anecdotes about Texas

*FREE on cassette

Note: Available to Texas stations only

Longhorn Radio Network: 800-457-6576

LIKE IT IS

29-minute, weekly interview program with a focus on the Afro-American
community

$5.00 per program on cassette (plus ordering fee)

Longhorn Radio Network: 800-457-6576

LIVE! AT THE CONCERTGEBOUW

120-minute, weekly program of classical music concert performances (13
weeks) (+)

*FREE on DAT. Contact producer for shipping fees.

WCLV: 800-491-8863, Attention: Tara Renk

LIVE! FROM ROTTERDAM

120-minute, weekly program of classical music concert performances (13
weeks) (+)

*FREE on DAT. Contact producer for shipping fees.

WCLV: 800-491-8863, Attention: Tara Renk

LIVING WITHOUT LIMITS

29-minute, weekly program focusing on the needs and interests of
Americans with disabilities and chronic illnesses

$5.00 per program on cassette (plus ordering fee)

Longhorn Radio Network: 800-457-6576

L'ORCHESTRE SYMPHONIQUE DE MONTREAL

120-minute classical music concert performance (13 weeks per year) (+)

On 10.5" open reels or DAT—tape return. Contact producer for fees.

WFMT Fine Arts Network: 773-279-2110

MAKING CONTACT

29-minute, weekly public affairs program of grassroots voices and ideas with stories of change through activism

Excerptable. Contact producer for local underwriting restrictions.

*FREE on cassette (+)

National Radio Project: 415-851-7256

MASTER CONTROL

28-minute, weekly magazine interview program with inspirational commentary

*FREE on CD

Media Technology Group/Southern Baptist Convention: 800-266-1837

ME AND MARIO

29-minute, weekly political discussion program featuring former New York governor Mario Cuomo. Program focuses on national political issues.

(+) Produced by WAMC-Albany

*FREE on cassette. DAT available. Contact producer for details.

National Productions: 800-323-9262, ext. 165

MEDIA NETWORK

30-minute, weekly program on developments in electronic media, with an emphasis on international broadcasting, high-fidelity, video, and computers.

*FREE on cassette

Radio Netherlands: 800-797-1670

MICROLOGOUS

30-minute, weekly program of early music (+)

*FREE on DAT. Contact producer for shipping fees.

WCLV: 800-491-8863, Attention: Tara Renk

MIDNIGHT SPECIAL

60- or 120-minute, weekly program of folk music, comedy, and show tunes (+)

*FREE on 10.5" open reels or DAT—tape return. Stations may take first hour or both hours.

WFMT Fine Arts Network: 800-USA-WFMT ext.2112

MIDWEST TODAY RADIO EDITION

5-minute, weekly hard news or feature module from the contents of *Midwest Today* magazine. Module features exclusive interview segments from the magazine's profiled celebrities and newsmakers as well as reports from the magazine's columnists and reports on events in the Midwest.

*FREE on CD delivered every six weeks. CD contains :30 taggable promo for each program. Program contains incidental reference to the magazine's telephone number (deletable by noncommercial stations).
Midwest Today
Radio Edition: 515-755-3851

MILLENNIUM OF MUSIC
59-minute, weekly program of early music. (Produced by WETA-Washington, DC) (+)
*FREE on DAT. Contact producer for shipping fees.
WCLV: 800-491-8863, Attention: Tara Renk

MILWAUKEE SYMPHONY ORCHESTRA
120-minute, classical music concert performance (13 weeks per year). (+)
On 10.5" open reels or DAT—tape return. Contact producer for fees.
WFMT Fine Arts Network: 773-279-2110

MIND'S EYE AUDIO PRODUCTIONS
Various 59-minute, magazine-style documentaries built round a topic or theme. Each program features interviews, music, readings, and dramatizations. (Also available on Pacifica Satellite System)
On cassette or DAT. Contact producer for details.
Mind's Eye Audio Productions: 608-256-6525 or *mindseye@itis.com*

MOZARTWOCHE
120-minute, weekly program of classical music performances/highlights from the week-long Mozart festival in Salzburg (+)
*FREE on DAT. Contact producer for shipping fees.
WCLV: 800-491-8863, Attention: Tara Renk

MUSIC AND THE SPOKEN WORD
29-minute, weekly program of music by the Mormon Tabernacle Choir along with short (~2 minutes) spoken message on values, etc.
*FREE on CD, no excerpting allowed
Marilyn Throckmorton/Bonneville Communications, 800-247-6655

MUSIC FROM AUSTRIA'S IMPERIAL CITIES
120-minute, weekly program of classical music concert performances (26 weeks) (+)
*FREE on DAT. Contact producer for shipping fees.
WCLV: 800-491-8863, Attention: Tara Renk

MY MUSIC
30-minute, weekly panel, quiz program on music from classical to popular (Produced by the BBC)
On reels, DAT, cassette—tape return. (+)
Contact producer for fees.
WFMT Fine Arts Network: 773-279-2110

MY WORD
> 30-minute, weekly panel, quiz program on language and wordplay (Produced by the BBC) (+)
> On reels, DAT, cassette—tape return.
> Contact producer for fees.
> WFMT Fine Arts Network: 773-279-2110

NASDAQ STOCK MARKET REPORT
> 30- and 60-second, weekday stock market reports with business news, updated throughout the day
> *FREE by telephone: 800-536-0837
> Cameron Brown: 202-728-8379

NATIONAL INSTITUTES OF HEALTH RADIO NEWS SERVICE
> 60-second, weekly news package on health and medical issues
> *FREE by telephone: 800-MED-DIAL
> Available 5am-5pm ET weekdays
> NIH Broadcast Services: 301-496-5895

NATIONAL PRODUCTIONS SAMPLER
> 29-minute, weekly magazine program of news modules on the environment, women's issues, legal affairs, health, education, literature, and national politics. Segments are from other WAMC productions (The Environment Show, The Law Show, The Best of Our Knowledge, The Health Show, etc.). (+) Produced by WAMC-Albany.
> *FREE on cassette. DAT available. Contact producer for details.
> National Productions: 800-323-9262, ext. 165

NATURAL GAS RADIO NEWSLINE
> Weekly news feed of energy news and features
> *FREE by telephone: 800-336-4795
> American Gas Association: 703-841-8667

NEWSBREAKS FROM BRITAIN
> 5 to 7 weekday voice wraps on European news
> *FREE by telephone (800-762-7667) or ISDN (212-754-9415)
> British Information Services: 212-745-0376

NEWSCAST FROM THE CHURCH OF JESUS CHRIST OF LATTER-DAY SAINTS
> 15-minute, weekly newsmagazine of news and features of the LDS church
> *FREE on cassette
> Church of Jesus Christ of Latter-day Saints: 801-240-4612

NORTHWESTERN NEWSFEED
> 10 voice wraps/thrice weekly of news and information from Northwestern University
> *FREE by telephone: 800-942-1145
> Northwestern University Broadcast Relations: 847-491-5753

NUVEEN/LYRIC OPERA OF CHICAGO
Variable-length opera performances (8 weeks per year) (+)
On 10.5" open reels or DAT—tape return.
Contact producer for fees.
WFMT Fine Arts Network: 773-279-2110

PACIFICA NETWORK NEWS
29-minute, weekday newsmagazine of world and national news
Excerptable by telephone (+)
Sliding fee schedule based on station budget. Contact producer for fees.
NOTE: Pacifica is replacing phone feeds with new Ku band satellite service.
    Contact producer for details.
Pacifica Network News: 202-588-0988

PERSPECTIVE
15-minute, weekly program analyzing a specific political, economic, social
    or humanitarian issue
*FREE on cassette
United Nations Radio: 212-963-6982

PHILADELPHIA ORCHESTRA
120-minute, weekly classical music concert performances
In development. Contact producer for details.
WFMT Fine Arts Network: 773-279-2110

POWERLINE
28-minute, weekly program of adult contemporary music of the '70s, '80s,
    and '90s along with brief inspirational messages
*FREE on CD
Media Technology Group/Southern Baptist Convention: 800-266-1837

PROGRESSIVE MEDIA PROJECT COMMENTARIES
2- to 3-minute commentaries on multicultural and progressive issues (5-8
    per month)
*FREE on cassette
Progressive Media Project/Catherine Capellaro: 608-257-4626

PUBLIC BROADCASTING RADIO SERVICE
Various drama, children's, comedy, documentary and music programs,
    including selections from the Canadian Broadcasting Corporation.
On CD. Contact distributor for fees and underwriting information.
Public Broadcasting Radio Service: 781-595-3747

RACHEL'S PLACE
29-minute, weekday program of classic fairy tales and children's stories.
    One program per week is stand alone for stations desiring a weekly
    rather than daily service. (+) Produced by WAMC-Albany.
*FREE on cassette. DAT available. Contact producer for details.
National Productions: 800-323-9262, ext. 165

RADIO NATION

29-minute, weekly newsmagazine program featuring interviews and news relevant to the progressive community on cassette. Contact producer for fees.

The Nation: 212-209-5400

RADIO NETHERLANDS DOCUMENTARY

30-minute, weekly documentaries on subjects of worldwide interest
*FREE on CD (CD includes EUROQUEST)

Radio Netherlands: 800-797-1670

REFUGEE VOICES

Thrice-yearly program of four feature modules (3 to 4 minutes each) with refugees and/or their advocates discussing the refugee experience
*FREE on cassette or reel

Refugee Voices: 800-688-7338

ROCK AND RAP CONFIDENTIAL REPORT

15-minute, quarterly program of current music and 2 to 3 political topics per program
*FREE on cassette

Rock and Rap Confidential: 310-398-4477

SALZBURG MUSIC FESTIVAL

120-minute, weekly program of classical music concert performances (13 weeks) (+)
*FREE on DAT. Contact producer for shipping fees.

WCLV: 800-491-8863, Attention: Tara Renk

SAN FRANCISCO SYMPHONY

120-minute, weekly program of classical music concert performances (26 weeks) (+)
*FREE on DAT. Contact producer for shipping fees.

WCLV: 800-491-8863, Attention: Tara Renk

SCIENCE REPORT RADIO

2-minute, module about science. 80 programs per year delivered quarterly on CD (20 programs per CD).
*FREE on CD

American Institute of Physics: 301-209-3090, *pubinfo@aip.org*

SCOPE

15-minute, weekly magazine program covering aspects of the work of the United Nations
*FREE on cassette

United Nations Radio: 212-963-6982

SECOND OPINION

30-minute, weekly program with *Progressive Magazine* editor Matthew Rothschild interviewing leading writers, activists, and performers (+)

*FREE on cassette

The Progressive: 608-257-4626

SIDETRAX

60-minute, weekly program of acoustic, folk, bluegrass, and eclectic music built around a weekly theme

$7.50 per program on cassette (plus ordering fee)

Longhorn Radio Network: 800-457-6576

SISTERS AND FRIENDS

29-minute, weekly program featuring lifestyles, success stories, problems and dilemmas facing women of color

$5.00 per program on cassette (plus ordering fee)

Longhorn Radio Network: 800-457-6576

SOPHISTICATED SWING

60-minute, program of big band and swing music recordings

On CD (13 programs in series). Contact distributor for fees.

Public Broadcasting Radio Service: 1-781-595-3747

STAR DATE

2-minute, daily program about astronomy

$563 per year first year, $724 per year thereafter on monthly CD

Market exclusivity, excerpting allowed with permission from producer

University of Texas McDonald Observatory: Suzanne Harm at 512-475-6760

SWISS MIX

90-minute, monthly magazine of culture, medicine, politics, the environment, science, tourism, and human interest

Excerptable with credit to Swiss Radio International

*FREE on cassette

Swiss Radio International: *Robert.Zanotti@SRI.SRG.SSR.CH*

TENT SHOW RADIO

55-minute, weekly program of live stage shows of folk, jazz, bluegrass and acoustic music from Lake Superior's Big Top Chatauqua in Bayfield, Wisconsin. 26 programs per year, with stations allowed to repeat programs once for full-year service. No excerpting.

*FREE on CD (CD also contains program promo)

Phillip Anich: 888-BIG TENT

THIS WAY OUT

30-minute, weekly magazine program of lesbian/gay/bisexual issues with news update segment, feature stories, interviews, readings, and music segments (+)

$20 per month on cassette—billed quarterly in advance

Excerpting allowed with credit

Greg Gordon: 818-986-4106

TIME AND SEASONS FAMILY EDITION

30-minute program of family and marriage issues (5 programs in series)

*FREE on cassette

Church of Jesus Christ of Latter-day Saints: 801-240-4612

TRAVELER'S JOURNAL

2-minute, weekday modules highlighting the joys of travel and practical advice (+)

*FREE on cassette, DAT, ISDN, Internet

(*www.travelersjournal.com*) Stations must complete broadcast agreement and air segments in their entirety. No market exclusivity.

David Bear: 800-860-7537

TRENDS IN REAL ESTATE

90-second, weekly report on real estate, home buying, mortgages

*FREE by telephone: 800-832-0338

National Association of Realtors: 202-383-1177

UK BLUES

30-minute program of blues recordings, performances and interviews from the United Kingdom

On CD (26 programs in series). Contact distributor for fees.

Public Broadcasting Radio Service: 781-595-3747

UN AFRICA

30-minute, monthly magazine program of United Nations activities of specific interest to Africa

*FREE on cassette

United Nations Radio: 212-963-6982

UN CALLING ASIA

15-minute, weekly magazine program of United Nations activities of particular interest to Asia and the Pacific

*FREE on cassette

United Nations Radio: 212-963-6982

UN CARIBBEAN ECHO

15-minute documentary program on a major United Nations issue of special
    interest to the Caribbean countries
*FREE on cassette
United Nations Radio: 212-963-6982

UN CARIBBEAN MAGAZINE

15-minute, weekly modular newsmagazine program of United Nations
    activities of particular interest to Caribbean countries
*FREE on cassette
United Nations Radio: 212-963-6982

U.S. DEPARTMENT OF AGRICULTURE

Daily telephone newsline
*FREE via telephone
30-minute weekly program of agriculture topics
*FREE on cassette
USDA: 202-720-6072

U.S. DEPARTMENT OF TRANSPORTATION

Daily news packages on transportation news
*FREE via telephone: 800-526-1144
USDOT: 202-366-5565

UNIVERSITY OF MINNESOTA NEWSLINE

Daily news packages available by telephone
*FREE via telephone: 612-625-6666 or via Internet at *www.umn.edu/
    urelate/audio/*
Note: regular cassette of produced news segments to be produced in the near
    future. Contact producer for details.
J. B. Eckert: 612-624-5228

UNIVERSO

2-minute, daily program on astronomy and space science in Spanish
*FREE on CD
University of Texas, McDonald Observatory: 512-475-8843

VAN CLIBURN INTERNATIONAL PIANO COMPETITION

59-minute, weekly program of classical music piano recitals, performances
    (26 weeks) (+)
*FREE on DAT. Contact producer for shipping fees.
WCLV: 800-491-8863, Attention: Tara Renk

VIENNA PHILHARMONIC

120-minute, weekly program of classical music performances (13 weeks)
    (+)
*FREE on DAT. Contact producer for shipping fees.
WCLV: 800-491-8863, Attention: Tara Renk

THE VOCAL SCENE
54-minute, weekly hosted program of opera recordings (+)
$25.00 per program on tape, tape return basis
Radio Features Corporation: 301-718-2800

WEATHER NOTEBOOK
2-minute, weekday module on weather and weather lore. (+)
    Coproduced by New Hampshire Public Radio.
*FREE on cassette. Tape includes weekly program promo.
Mount Washington Observatory: 603-356-2137

WEEKEND RADIO WITH ROBERT CONRAD
59-minute, weekly program of classical recordings with crossover material
    and comedy selections (+)
*FREE on DAT. Contact producer for shipping fees.
WCLV: 800-491-8863, Attention: Tara Renk

WE'RE SCIENCE
29-minute weekly program answering questions submitted by listeners about
    science and technology. (+) (Produced by KUMR-Rolla, Missouri)
*FREE on cassette
Must air within 30 days of receipt. No excerpting.
John Francis: 800-327-6440

WHEN THREADS COME LOOSE
30-minute, radio dramas produced, directed, acted by students (39 in series)
    (Produced by KUOM-AM Minneapolis)
*FREE on cassette
 Must run entire series and contact producer with broadcast times.
Andy Marlow c/o KUOM-AM: 612-625-3500

WOMEN
15-minute, weekly program examining issues affecting women around the
    world
*FREE on cassette
United Nations Radio: 212-963-6982

WOMEN'S INTERNATIONAL NEWS GATHERING SERVICE (WINGS)
Two programs on women's issues and women's perspectives with regard to
    the environment, politics, law, labor, sexuality, and social change:
    30-minute monthly newsmagazine, newscast. Excerpting of modules
    permitted.
30-minute, thrice-monthly news/current affairs program with extended
    interviews, speeches, or documentaries.
$99 per quarter for both programs on cassette

$90 per year for monthly newscast only on cassette (contact producer about subsidized subscriptions)
WINGS: 800-798-9703/*wings@wings.org*

WORD FOR THE WISE
2-minute module. 20 per month on language and word origins (2 per week are undated and usable on weekends). (+) Produced by WAMC-Albany
*FREE on cassette. DAT available. Contact producer for details.
National Productions: 800-323-9262, ext. 165

WORLD BUSINESS REPORT
90-second, weekday news report on business and finance from the resources of Merrill Lynch
*FREE via telephone: 800-626-6750
North American Network: 301-654-9810

WORLD CHRONICLE
30-minute, weekly panel discussion program featuring international media correspondents questioning United Nations officials about global issues (39 programs per year)
*FREE on cassette
United Nations Radio: 212-963-6982

THE WORLD IN REVIEW
15-minute, weekly newsmagazine program covering current events at the United Nations
*FREE on cassette
United Nations Radio: 212-963-6982

ZBS PRODUCTIONS
Variable-length full-cast audio dramas, multipart series on cassette
Contact producer for fees and information
ZBS Foundation: 800-662-3345

B

Sample Study

The following provides details of an actual study of radio listening habits among the residents of a college town conducted by the author during his master's degree studies. It is presented as an example of self-studies that can be carried out by or for the college radio station on a local basis. Included is a concise history of college radio in the United States to establish need and viability of the medium, a brief history of the station itself, and the hypothesis findings of the study. The study was performed in 1979.

HISTORY OF COLLEGE RADIO (UP TO 1979)

For the majority of its life FM radio has lagged behind AM in economics and audience. While AM radio got its start in the early 1920s, the first patents for FM radio were not filed until 1933. In 1938 there was still only one FM station licensed to its inventor, Edwin H. Armstrong (Barnouw 1975, 78-83). In 1940 the Federal Communications Commission assigned the 42- to 50-megacycle band to FM with 40 possible channels (Lessing 1956, 193-94).

Because of the vast commercial success and popularity of AM radio, many individuals wanted to gain a foothold in the industry with an FM radio station. By 1941, 67 stations were licensed. But with the coming of the war and subsequent freezing of radio set production, 21 of these stations had left the air by the end of the war (Sterling 1968).

In 1945 the FCC dealt a shattering blow to the progress of FM radio when it shifted FM assignments to the 88 to 106-megacycle band (U.S. Federal Communications 1945, 20-21). This immediately made over half a million radio sets for FM obsolete (Lessing 1956, 258-59). While in the long-run the shift benefited FM radio, it was a severe blow to the just-developing industry.

Following the war, more than 400 applications for FM stations were filed (U.S. Federal Communications 1945, 19). However, because many FM station owners also owned AM stations, the economics of running both was very high. As early as 1946, FM stations were simulcast with AM stations, sometimes as much as 100 percent of the program content (Sterling 1968).

In 1950, FM licenses began to decline and stations began to leave the air as the competition of AM radio and the developing television proved too much for FM. The low point came in 1956. In 1957 the number of FM stations began to increase again because of better and more receivers, a larger potential audience, the crowdedness of the AM band, and the interest in high-fidelity music with stereo recordings. In 1961 the introduction of stereo multiplex FM stations added another boost to FM radio (Sterling 1968).

In 1965 the FCC ruling that FM and AM stations could not simulcast 100 percent of their programs (which was further decreased to 50 percent and then to 25 percent in subsequent years depending on the size of the market covered), hastened a trend toward diversification among FM stations in music and other formats. This ruling effectively increased the number of radio stations with different programming by one-third in many of the large markets in the United States.

In 1970 FM stations were for the first time competing successfully with AM stations. By 1976 FM stations were getting 40 percent of the audience in most markets in the U.S., and were getting a larger share of the radio audience than AM stations in two markets—Washington, D.C., and Dallas-Ft. Worth.

However, this trend in the growth and popularity of FM stations has not carried over to the noncommercial educational FM stations. As commercial radio grew from 1928 to 1948, noncommercial, educational stations declined in numbers, abandoning their original AM assignments to commercial interests. This confirmed what many persons had thought earlier—that a share of AM frequencies should have been set aside exclusively for educational use. This was because educational interests could not be expected to compete reasonably with commercial interests in the open market (Head 1976, 150-52).

When Congress revised the Radio Act of 1927, this issue was revived and a proposal was made to reserve 25 percent of the AM channels for educational interests. But this proposal was not made part of the Communications Act of 1934 because many commercial licenses would have to have been revoked, as few desirable assignments were left in the AM band (Head 1976, 150-52).

Educational interests settled for a compromise—the FCC was to report to Congress on the advisability of allocating fixed percentages of radio channels to nonprofit radio programs. To their disappointment, the FCC in 1935 reported to Congress that commercial stations gave ample time to educational programming without requiring the allocation of radio channels to educational interests (Head 1976, 150-52).

However, the realization of the actual lack of such educational programs on commercial stations caused the FCC to reverse itself when allocating FM channels. In 1940, the FCC set aside five of the 40 channels in the FM band for education. This was not a very bold gesture at that time, since FM still had very small audiences and there was a limited demand for FM licenses. In 1945, when the FCC shifted the FM broadcast band, 20 of the 100 FM channels were set aside for educational interests (88 to 92 megacycles). These instances established the principle of allocation for noncommercial, educational interests, which was subsequently followed with allocation of television channels for the same purpose (Head 1976, 150-52).

To stimulate the use of FM frequencies, the FCC in 1948 liberalized its rules permitting the informal operation of 10-watt noncommercial FM radio stations. This was necessary, as many educational organizations did not have the money to start a larger operation than this. Many schools took advantage of this, and about half of the 615 FM educational stations on the air in 1974 were 10-watters (U.S. Federal Communications 1974, 64).

But the 10-watt rules also invited schools to start educational FM stations with no serious commitments, serving only as practice areas with no real attempt to undertake serious program service. Their presence became an embarrassment in the 1970s, when public broadcasting began to become a truly alternative service. It was difficult to find unassigned channels for full-powered noncommercial stations to use (Robertson and Yokom, 1973, 107).

Although FM audiences and revenue have increased vastly in the last 10 years, this has not been true for noncommercial, educational FM stations. But these stations were given a boost when, in 1967, they were made eligible for federal funds for construction and program aid from the Corporation for Public Broadcasting (Head 1976, 158).

A study in 1950 showed that noncommercial, educational radio embraced such a variety of enterprises with much-varying goals that it had no meaning for the general public. In 1971 to 1972 a 15-month tour and study of educational radio stations concluded that no stations were alike and there were almost no models to point to (Robertson and Yokom 1973, 115).

(The preceding historical information was adapted from the 1977 survey of North Texas State University students conducted by Laurie Fries Andrews.)

This study concerned a noncommercial educational FM radio station licensed to North Texas State University in Denton, Texas, conducted in the spring of 1979 and reported on May 1, 1979.

BACKGROUND (UP TO 1979)

KNTU(FM) is licensed to North Texas State University (now the University of North Texas) in Denton, Texas. It began operation in November 1969 (actually, the first broadcast was Halloween, October 31, 1969) and operated with 440 watts of power at a frequency of 88.5 megahertz at the time of the study. It had a maximum coverage area of 20 miles under excellent reception conditions, and a coverage area of 6.5 miles under any conditions. KNTU shared its frequency with a Dallas radio station, KRSM, licensed to St. Mark's Academy.

PROBLEM

The problem of the study was to determine the radio listening habits of the residents of Denton and how the habits affected the radio listenership of KNTU by the Denton residents.

ANALYSIS OF FINDINGS

(The study methodology utilized was that of a telephone coincidental recall employing a random sample from the city telephone directory. The survey instrument was pretested via a pilot study of the questionnaire.)

The results of the questionnaires were analyzed in five parts.
1. Comparison of the demographic characteristics of the Denton residents of the analyzed questionnaires.
2. Breakdown of the general AM and FM radio listening habits of Denton residents concerning the number of sets they had access to daily, the type of radio they usually listened to, the number of hours spent listening to AM and FM radio, and the times of the day spent listening to AM and FM radio.
3. Preferences of Denton residents toward specific AM and FM radio stations, the number of AM and FM stations they listened to, and preferences in music and nonmusic formats.
4. Breakdowns of specific listening habits of Denton residents to KNTU concerning the time of day listened, and specific questions concerning the programs offered by KNTU.
5. Breakdown of the specific listening habits of the Denton residents to KNTU concerning KNTU's limited broadcast range due to its (then) current power output.

DISCUSSION

The three hypotheses set forth for this study were all rejected based on a lack of data to connect the facts uncovered with reason for their occurrence.

The first hypothesis stated the Denton population listened consistently to one or two radio stations because of the stations' homogeneous format. (Homogeneous format was defined as consistent adherence to a particular type of music and limited programming besides music that is also adhered to strictly. News format stations could also fall into this category.) Although the top two AM and FM stations listened to by Denton residents accounted for 38 and 25 percent respectively, the findings presented disproved this hypothesis. No relationship could be drawn concerning the reason or reasons for this pattern from the data in this study. Because the questionnaire did not ask for the type of program schedule, whether homogeneous in music and other programs, the relationship indicated in the hypothesis could not be supported.

The second hypothesis stated that if a radio station had a very heterogeneous format aimed at many segments of the general population, it would not reach a majority of any of the different audiences with the general population. (This hypothesis was set forth because of KNTU's very heterogeneous format of music and programs in comparison with other radio stations. While most radio stations played one type of music, KNTU had many programs of different types of music as well as a wide variety of other special-interest programming aimed at various specific groups.) Although it was found that KNTU, with its heterogeneous format, was listened to regularly by 50.54 percent of the respondents, this was not listed by the residents as a reason for not listening to KNTU. By looking at the findings, it could be seen that only 5.77 percent of the respondents listed KNTU as one of their preferred FM stations; therefore, there was not a correlation between the two groups that could be made to support the hypothesis.

The third hypothesis stated that KNTU's limited broadcast range due to its then current power output had a negative effect on the amount of KNTU's audience. Although over half of the total survey group said they might or would definitely listen to KNTU, or listen to it more if they could receive it outside of Denton, only 68 percent of the respondents answered this question. Due to this fact, no correlation could be drawn between the negative and positive feelings of the group.

It was found that since this study is primarily a fact-finding one, relationships such as were suggested in the hypotheses could not be supported by the data obtained.

This analysis had only percentage breakdowns to the answers of the various questions. It is hoped that the questionnaires will be further broken down using a Spearman correlation coefficient test and chi-square tests in the future in order to uncover more findings in the results.

Unfortunately, the lack of time was a major factor in the analysis of the findings in this survey. (Additionally, as with any nonprofessional study, it should be cautioned that listener projections are at times difficult to predict and can lack accuracy. Keeping in mind such limitations, college stations can use such apprentice studies to help provide an "overall" view of their performance.) (Source: Sauls 1979).

(Additionally, see Marilyn L. Boemer's 1990 *Feedback* article, which details audience surveys also carried out for KNTU-FM and Channel 36, the cable-access channel for then North Texas State University.)

REFERENCES

Barnouw, E. (1975). *Tube of plenty: The evolution of American television.* New York: Oxford University Press.

Boemer, M. L. (1990). University station audience surveys: Learning research methods. *Feedback* 31(3): 14-16.

Head, S. W. (1976). *Broadcasting in America.* Boston: Houghton Mifflin.

Lessing, L. (1956). *Man of high fidelity, Edwin H. Armstrong.* Philadelphia: J. B. Lippincott.

Robertson, J., and G. G. Yokom. (1973). Educational radio: The fifty-year-old adolescent. *Educational Broadcasting Review,* April, pp. 107-16.

Sauls, S. J. (1979). *A study of radio listening habits of the residents of Denton, Texas.* Denton: KNTU-FM.

Sterling, C. H. (1968). WTKJ-FM: A case study in development of FM broadcasting. *Journal of Broadcasting* 12(4): 341-52.

U.S. Federal Communications. (1945). *Eleventh annual report.* Washington, D.C.: Government Printing Office.

U.S. Federal Communications. (1974). *FCC Annual Report.* Washington, D.C.: Government Printing Office.

Creating a World Wide Web Site for a College Radio Station: Considerations, Concerns, and Strategies

Craig A. Stark

Sam Houston State University

(Paper presented at the Texas Association of
Broadcast Educators Conference, March 1, 1997,
in Dallas, Texas. Used with permission.)

The World Wide Web is an ever-increasing, always developing source of information that could very soon become one of the most popular methods of communication. By combining the web with other Internet services such as email, many people believe that the mass use of electronic communication such as the web is unavoidable. The web can hold a wealth of information for persons studying mass communications. Web sites with radio, television, and film themes are commonplace and growing at a very quick rate. Many colleges and universities are also getting in on the act, not by just creating web sites for the school itself and many related activities, but also by creating sites for their campus radio and/or television stations. When it comes to creating a web site for your college station, there are several things that should be kept in mind. The main thing to remember is that a web site produced for a non-commercial college or university station is different than one produced for a commercial station. The differences mainly deal with the actual cost of producing and maintaining the web site, along with the general purpose of the site. I would argue that one of the main purposes of a university web site of any kind would be to educate and inform, whereas the purpose of a commercial site would be primarily to entertain and make a profit. In the following paragraphs, I will discuss the development of the web site for KSHU, Sam Houston State University's campus radio station, and discuss several key strategies which faculty and/or staff administrators may want to consider when it comes to topics such as security, content, and promotions. One word of note: since language and jargon in the computer world seem to be changing all the time, it can become very difficult to distinguish between the "Internet" and the "web" and a "site" versus a "page". These four terms are used almost interchangeably on a daily basis. With that in mind, please note that I do indeed use these words interchangeably throughout this paper.

WHY SHOULD I EVEN BUILD ONE OF THESE THINGS?

The Internet is fast becoming one of the most widely used methods of communication in the world. It should be fairly obvious that use of the Internet has increased over the past several years. With many more homes and families purchasing personal computers equipped with modems, computers could soon become as common in the average home as microwave ovens or refrigerators. Many experts believe that access to the Internet can only increase as time goes by. With the development of the new Web TV, families and households do not even need a computer to gain access. This system works with just a modified television set and a remote control. With advancements in technology such as these, it is quite arguable that use of the Internet will only increase in the coming years.

Many companies and organizations use the Internet as a means of getting their message across to potentially millions of people. Colleges and universities also use the web as a means of informing the general public about their services and programs, as well as a method of recruiting potential students. With aspects such as these in mind, it is important to distinguish the differences between a web site created for commercial purposes and a web site created for noncommercial purposes. In the case of a commercial entity (radio station business, etc.), the main purpose is to make a profit. Corporations and other commercial businesses use the web to increase knowledge about their product or service, which will hopefully lead to increased consumer interest and therefore more sales. A noncommercial entity (college radio station, charity organization, etc.) may also try to generate interest so that financial revenue is increased. In my opinion, however, a noncommercial entity has a much greater opportunity and responsibility to also work in several other areas. These areas include recruiting, general and specific information about the entity, public service, any pertinent local information, and the opportunity to learn more about the particular industry that the entity is associated with. I believe that a nonprofit organization can benefit greatly from using the Internet if these opportunities and responsibilities are fulfilled. It was mainly for these reasons that I began working on the KSHU web site.

HOW DO YOU GET STARTED?

One of the easier ways to at least begin developing a web site for your station is by purchasing or acquiring a rather inexpensive web builder program. Many of these programs can be purchased in computer software stores for a nominal fee, acquired from the computer services department at your university, or acquired via shareware from the World Wide Web. The main advantage to using a builder or start-up program is that it gets you up and running in a relatively short period of time. Web sites are programmed in HTML. This stands for HyperText Markup Language. It is basically a computer program that allows you to create web sites. If you are unfamiliar with HTML, start-up programs like those mentioned allow you to get going and to get a grasp as to what your site will look like. They are also good confidence boosters! If you are unsure of your ability to create something such as a web site, you can see the results of your endeavors almost immediately. For the KSHU web page, I used *Quick Start,* a simple, easy to use program that asks for information to be placed on the web page and then automatically translates it into HTML. Your web site is displayed through the web browser installed on your computer. Once I had the web site built on my home computer, it was a simple process to copy it to a floppy disk and take it to computer services, where they installed it on the university's system. I found that once the web site was in place on the school's system, it had to be edited by programming in HTML. Honestly, this appeared to be a daunting process at first, but with time and effort, it became rather easy to learn. I have found that there are several basic commands and rules one must know when programming in HTML. Once you learn these commands and rules, however, the remainder of the process is fairly simple. These basic commands are listed in Appendix C at the end of the paper.

When it comes to getting help, Computer Services at the university level have proven to be invaluable. As mentioned before in the case of KSHU, they were initially used for launching the web site and putting it in the system. Beyond that, however, they have been a tremendous aid in putting it all together and keeping it all together. Normally, the university Computer Services Department can help you maintain the site, offer suggestions on improving it, and even offer to scan photographs and graphics into digital images so that they may be placed on the web site. The nice aspect of all this is that they normally do not charge a fee for these services.

Another excellent source for information and assistance rests in the students themselves. Let's face it—many students have much more computer savvy these days than most of us do. I remember being proud of myself when I could program in Basic and COBOL. Even if students cannot help you with the programming of the web site, they can help you by pointing you in the right direction to get support, find graphics, find links to other web sites, etc. The students at Sam Houston State have been extremely helpful towards my efforts to build a web site for KSHU. They have offered information on links, where to find graphics and backgrounds, and have even provided me with a program to convert bit map images (normally from a Windows Paintbox) into .JPG (Joint Photographic Experts Group) and GIF (Graphics Interchange Format) files. The students are an excellent source of information and support.

On that note, there are also many sites on the web that will help you create the look you want. These sites provide backgrounds, graphics, and information on building a site. The addresses of several of these sites are located in Appendix B.

WHO'S IN CHARGE HERE?

Something that should be kept in mind while developing the web site is, who should be in charge of developing, protecting, and supervising the content of the site? As a teacher or administrator, you basically have three options: (1) allow student managers to maintain the web and hope they can be responsible enough with the content, (2) work on it yourself, or (3) hire outside help to maintain the site. This option may not be acceptable, however, especially if the outside help costs your department or organization more money. It may be possible for the Computer Services department at your university to maintain the page on their own, although you may run the same risk as mentioned above in the first option. A lot of what you decide may hinge on such factors as time, amount of computer knowledge, and patience. Due to past experience, I have found that it is a better idea for the Station Manager to have control over the web site and the information that is displayed on it. To be honest, the version of the KSHU web page that is attached to this paper is really the second version. [The book author recommends that you visit the KSHU web site at *http://www.shsu.edu/~rtf_kshu.*] The first version was created solely by students at the station in the spring of 1996, and maintained exclusively by them. As the Station Manager, I did not even have the passwords to change the content of the page. Needless to say, although I admired their effort, their end result was less than encouraging. When it was decided to revamp the page that summer, I pulled "executive authority" and requested that Computer Services on campus shut down the page and remove it from the system. I asked them to set up a new account and a new address. When this was done, work was started on the new web page. To be on the safe side, however, I saved a copy of the previous web page on floppy disk so the students would have a copy of their work. Many of them were not happy when they came back and saw what had happened. They had wanted complete control of the web site and felt that I was being rather authoritarian in taking it away. I told them that the page would not be taken away from them and that they still had access to it, but only from an outside viewpoint. If they had any suggestions, comments, or content to be placed on the site, I would be happy to hear them and work on them. Since then, I have been the only person who has worked on the web site and updated it on a regular basis. The point is, you need to look at what the situation is at your own university, in your own department, and decide what is the best for you. At KSHU, it was decided that the faculty would have control over the content and editing of the web site. The obvious advantage of this decision is to keep inappropriate material and content off the web page. Another advantage is that the web site maintains a steady, consistent "look," due to a faculty or staff member doing the upkeep and not a student who will be replaced by another student every six months or so. The rather odd thing I have noticed, however, is that although many students will not primarily concern themselves with the content of the site, they will make consistent recommendations on changing the look of the site, or some other superficial element. One complaint that a student had after the new site was up and running was that it looked too "booky." When I asked him to explain what he meant, he stated that there was too much to read on the site, even though everything on the site had pertinent information concerning the radio station and the university.

A disadvantage to maintaining the site yourself is that it just seems to add one more thing to the always-growing list of things to do. Web sites are always "under construction" and, in my opinion, should be updated at least once a week when possible. This can add more pressure and stress to an already busy job. I would suggest setting up a specific time during the week to work only on updating the web site. You may find it beneficial to try and work on it during the weekends when traffic on the university computer system may be lighter.

LINKS, LINKS AND MORE LINKS ...

What links should be included on the web site? To begin with, it may be important to know what a link is. A link is a connection from one web site to another, via the Internet. Normally, users will position their mouse arrow over a highlighted section of text, click on that section, and be automatically transported to another web site. This is a very easy and successful method of giving the user a greater opportunity to learn more about the organization and its associates. Creating a link to another site is normally an individual's call, and depends on what the goals of your web site are. At KSHU, I have found that links to several key areas can serve multiple goals. For instance, KSHU has a link to the Federal Communications Commission (FCC), the Southland Conference men's basketball site, and the Huntsville Chamber of Commerce, just to name a few.

The link to the FCC was an idea I got from viewing the University of Houston's web site. They had a link to the FCC, and I thought it was pretty interesting, so I included it. I think it's important to have a link to the FCC for two reasons. First, if a listener has a question or (gasp!) complaint about the station, they have a rather quick and easy way to reach the proper authorities. In my opinion, this helps promote proper operations of the station, shows the user that you are concerned about your on-air product, and helps you maintain a higher standard from your student staff. Second, it provides many students the opportunity to use your web site as an educational tool, and obtain information that could be used for academic purposes. It may be a good idea to accept the philosophy that your web site should educate and inform as well as entertain. I encourage my students in the Introduction to Broadcasting course at Sam Houston to use the KSHU web site for information to help them with their class assignments and papers.

The link to the Southland Conference may help members of KSHU's sports staff by providing them with information about the basketball team. It may be wise to consider modifying the site to accommodate your on-air staff. Several web sites even provide show prep for disc jockeys to use.

The link to the Huntsville Chamber of Commerce helps in several ways: (1) it allows for educational opportunities, (2) producers of news and sports at KSHU may find it useful for upcoming events and news, and (3) it can go a long way towards fostering a sense of goodwill with the local community. The producers of the Chamber of Commerce page were very happy when I asked if KSHU could make a link to their site, and agreed immediately.

Other links on your station's site could include a link to the university home page, a link to the department home page, if applicable, a link to a weather forecaster (even the Weather Channel is good for this), links to other specific university entities if possible (athletics, theater, music departments), links to label representatives for music information, links to news organizations and links to organizations such as the National Broadcasting Society and the Intercollegiate Broadcasting System. It is very important to let these organizations know that you are making a link to their sites, especially when these sites are produced on a local, or smaller scale. It just helps foster good relations. On the national or global scale, many of the larger web sites are not too concerned if you create a link or not. Many times, such as the case with the Chamber of Commerce, web publishers are more than willing to allow these links to be made. It is just another way to help them get their word out, and also help make you and your web page look better.

To assist in furthering the educational aspect of your site, create a link that sends the user to lists of radio stations across the country. When it comes to this, the best list I have found is the one at the Massachusetts Institute of Technology (address listed in Appendix A). Again, I encourage students in my classes to use this feature when it comes to some class assignments.

How do you find these sites to link to? You can purchase books or magazines that will help you find the place you are looking for. You can also ask others if they know of any sites that will help you. I have found, however, that the best way to find sites is to just get on the web and surf around. Use search engines such as *Yahoo* to find these sites. Not only will you probably find these sites faster and easier, you will also get a lot of experience using the web when you are looking around. This can help you quite a bit if you are somewhat inexperienced with using the web. Getting around on the web is a lot like anything else in life; the more you practice with it, the better you will become using it.

DOES ANYONE KNOW WE EXIST?

How do you promote the web site? That is, once the site is up and running, how do you let people know it is there? The web is a very large place, growing every day. It can be very easy to become lost amidst all the other information. Obviously, one of the best ways a radio station can promote its web site is by mentioning it on the air. Schedule liner cards for the disc jockeys to read, or produce at least one recorded promotional announcement that lets the listeners know the web page exists. Also, it may be helpful to enlist the aid of the campus newspaper or television station to promote the address. In many instances a promotional announcement in this medium can be exchanged for a promotional announcement on your station, or even a link on your web page. Even something as simple as putting the site address on the bottom of flyers or brochures about station activities can be beneficial. The important thing is to get the word out, and get it out in a way that will generate interest so that the listener will check out your site more often.

Another way to promote your web site is to have it included somewhere on the vast lists of radio stations that exist on the web, such as the MIT list mentioned above. Many label representatives and occasional surfers will go to these lists to help narrow their search. If you are listed here, it could be to your advantage. One word of caution: Some of these providers charge a fee to be listed. Watch out for this and don't pay unless you want to. To be honest, many of the free lists are just as effective as the commercial lists, and sometimes even more so (they are easier to find, take less time to download, etc.). In addition, many smaller regional or state organizations such as the Texas Association of Broadcasters, the National Broadcasting Society, and the Intercollegiate Broadcasting System will also provide listings for your station, and in many cases even provide direct links to your web site. Normally, all you have to do to put your station on a list is to access their site and fill out a short electronic form. Just be sure to give the correct information on your site address ("htpp" versus "http," etc.), and give the correct information on personnel, contacts, etc.

WHAT DOES THE FUTURE HOLD?

As mentioned before, web sites are continuously "under construction." They need to be constantly updated, modified, and worked on. Any site that you may develop cannot be left alone to "dry up and rot." Always look for new and better ways to improve the web site. In the long run it will pay off by (1) being more attractive to your listeners who surf in, (2) making the university and department look better, and (3) providing more educational opportunities for your students to use. On that note, some future plans for the KSHU web site include: Top Ten lists of music played on the station, promotional announcements about specialty programming and contests, more links to communication related entities, a link to Houston Metro Traffic Control and possibly a hookup to the Real Audio program, in order for KSHU to broadcast live on the web.

In the past several paragraphs I have detailed my personal experiences in creating a web site for a university radio station, and noted some of the lessons which were learned. Creating a web site can be a monumental task that takes a lot of time, energy, patience, and desire. With the right attitude, however, and learning from the experiences of others, it can become one of the more rewarding accomplishments of a person's career. Seeing something built from the ground up is an incredible thing to witness. Seeing it built from the ground up and recognizing it as your own work, however, is a wonderful experience which holds an extremely satisfying feeling.

APPENDIX A: LIST OF WORLD WIDE WEB SITES ON THE KSHU WEB PAGE
http://www.shsu.edu/~rtf_kshu

FEDERAL COMMUNICATIONS COMMISSION
http://www.fcc.gov

INTERCOLLEGIATE BROADCASTING SYSTEM
http://www.ibsradio.org

MAJOR LEAGUE BASEBALL
http://www.majorleaguebaseball.com

MIT LIST OF STATIONS ON THE WEB
http://wmbr.mit.edu/stations/list.html

NATIONAL BASKETBALL ASSOCIATION
http://www.nba.com

NATIONAL BROADCASTING SOCIETY
http://www.onu.edu/org/nbs

NATIONAL COLLEGIATE ATHLETIC
ASSOCIATION BASKETBALL
(with SLC link)
http://www.nando.net/SportServer/basketball/col.html

NATIONAL FOOTBALL LEAGUE
http://www.nfl.com

SAM HOUSTON RADIO/TELEVISION DEPARTMENT
http://www.shsu.edu/~rtf_www

TEXAS ASSOCIATION OF BROADCASTERS
http://www.tab.org

APPENDIX B: WEB SITE BUILDING HELP

ANIMATED GIFS: (animated pictures that can be used on your site)
http://members.gnn.com/dcreelma/imagesite/image.html

BOOKS:
Krol, E. (1995). *The Whole Internet User's Guide.* Sebastopol, CA: O'Reilly
& Associates.
Smith, B., Bebak, A. (1996). *Creating Web Pages for Dummies.* Foster City,
CA: IDG Books Worldwide, Inc.

ICONS: (non-animated pictures that can be used on your site)
http://www.acsu.buffalo.edu/Icons/icons.html
http://www.baylor.edu/icons/
http://www.geocities.com/SiliconValley/6603/

SEARCH ENGINES:
http://info-s. com
http://www.yahoo.com
http://werbach.com/web/web.html
http://www.hotbot.com/

APPENDIX C: HTML COMPLIANT TAGS

These are some of the basic commands that you must be able to work with, in
order for HTML to work at its best. Most of these tags may be used with any
type of browsers and/or software for programming in HTML. Remember,
HTML is case sensitive. Tag commands *must* be all capital letters.

TAG NAME	TAG	NOTES
Document type	<HTML></HTM>	Beginning & end of file
Title	<TITLE></TITLE>	Must be in header
Link	<A HREF></A>	Creates links
Images & Icons	<1MG SRC =>	Places pictures & graphics
Bold	<B></B>	Highlights text
Italics	<I></I>	Italicizes text
Headlines	<H1></H1> <H2></H2> <H3></H3>	Creates a large headline Creates a larger headline You get the idea...

[As stated earlier, at the time of writing of this book, you can view the current Web site for the campus radio station at Sam Houston State University at *http://www.shsu.edu/~rtf_kshu.*]

Index